RFID

RFID
Radio Frequency Identification

Steven Shepard

McGraw-Hill
New York • Chicago • San Francisco • Lisbon • London • Madrid
Mexico City • Milan • New Delhi • San Juan • Seoul
Singapore • Sydney • Toronto

The McGraw-Hill Companies

Cataloging-in-Publication Data is on file with the Library of Congress

1 2 3 4 5 6 7 8 9 0 DOC/DOC 0 1 0 9 8 7 6 5 4

ISBN 0-07-144299-5

The sponsoring editor for this book was Steve Chapman and the production supervisor was Pamela Pelton. It was set in Fairfield by MacAllister Publishing Services, LLC.

Printed and bound by R R Donnelley

This book is printed on recycled, acid-free paper containing a minimum of 50 percent recycled de-inked fiber.

DEDICATION

This book is for Dennis McCooey with love and admiration: best friend, photography partner, diving buddy, and confidant. Thank you for always being there for me and my family, and for 25 years of adventures.

Este libro también lo dedico a mis amigos y colegas en España durante este periodo tan horroroso. Os mando todos un fuerte abrazo. Viva España, y viva la libertad y la paz—para todos.

CONTENTS

ACKNOWLEDGMENTS

This book turned out to be an adventure to create. When I write about the technology industry, I often find myself doing research deep inside central offices, data centers, and office buildings; in the murky depths of cable vaults, on the distal ends of bucket truck arms, and in underground chambers filled with spiders and several inches of water; and aboard cable-laying ships and atop fiber-pulling towers. This book took me to some new and unexpected places. The University of Vermont's veterinary school showed me how RFID devices help them track their dairy herd; the Barnes & Noble bookstore in Burlington, Vermont, showed me how their RFID-based anti-theft system works and helped me understand the challenges of implementing the system; and Chris Hannon took me into the remarkable world of pigeon-racing—which relies on RFID tags affixed to birds' legs to precisely measure the arrival time of racers. Ben Ritter of Infineon Corporation and members of the European-American Chamber of Commerce gave me a glimpse into the evolving and ever-more-critical world of container security, using RFID technology to safeguard the world's large shipping ports.

At Texas Instruments, Michael Stich helped me secure access to the techno-cognoscenti of RFID in his company, while Bill Allen, Christine Cayer, and Randy Culpepper buried my desk under an avalanche of every conceivable type of RFID device. Thanks to their kindness and generosity, photos of the dizzying array of RFID tags are scattered throughout the book. Thanks also to Kenn Pilczak and Matt Hubbard of Atmel, Clerk Béat from SOKYMAT, and Ashi Majid, Joerg Borchert, Frank Gillert, and Casey Gozzolino of Infineon for their assistance with product knowledge.

Of course, there's also the more mundane matter of reading, correcting, critiquing, and polishing the manuscript that ultimately became this book. As always, I relied on a dedicated cadre of friends and colleagues to intellectually shred the manuscript and help me reassemble it into a readable product. These are the people who add the richness to the text, hanging application examples, implementation anecdotes, and technology depth to my skeletal manuscript. These people include Phil Asmundson, Larry Kivett, Naresh Lakhanpal, and Toni Nelson of Deloitte; Jack Tongue and Jack Gerrish of Agere Systems; Barbara Jorge and Anthony Contino of Lucent Technologies; Paul Bedell of SBC; Richard Parlato, Bob Kaphan, and Bob Maurer of Proximity, Inc.; Gary Kessler and Dave Whitmore of Champlain College; Jack Garrett of Garrett Business Technology; Gary Martin of TM2, Inc.; Dennis McCooey of See Life Productions; and Joe Candido of Fifth Element Associates.

A special thanks to Kenn Sato who, once again, read this manuscript repeatedly, adding value each time.

At the University of Southern California's Center for Telecom Management, I am grateful to Phil Cashia and Morley Winograd who direct the Center's programs, and also to instructor Ron Hubert of Hozho Associates who teaches in the Advanced Management and Executive Leadership Programs at USC. I also wish to thank Marty O'Toole, Graeme Rankine, and Sundaresan Ram of Thunderbird University in Glendale for additional insights into the international aspects of RFID deployment.

My editor and friend Steve Chapman deserves special thanks. I enjoy working with a professional of any sort; a professional with insight and passion about the publishing industry is like a triple play. Thanks, Steve.

As always, I reserve this last paragraph for my family. You are my world—and a wonderful place it is.

INTRODUCTION

One hot summer day in late July, Dave Whitmore walks into his favorite consumer electronics store, intent on looking one more time at the expensive game console he has been trying to convince himself to buy for the last two months. As he enters the store, he walks through a stream of air that, unknown to him, directs a spray of microscopic particles onto the legs of his pants. The particles contain passive radio devices that are activated when they pass within a few feet of sensor devices hidden throughout the store. When activated, they broadcast a simple identification message that is captured by each sensing device. The sensors, in turn, pass the identification information to a central database that analyzes the movement of the individual particles as they are carried around the store by their unwitting host, like a hitchhiking seedpod on a cat.

Dave walks down the game cartridge aisle on his way to the consoles, pausing to examine a few games along the way. He picks up one game several times before moving on. He stops briefly at other displays in the store that catch his interest before finally arriving at the console display.

Dave has been in this store several times to see the product that interests him, and each time he has left empty handed. This time he may buy: He *really* wants the game. He picks up the box, reads the information printed on the back as he has at least 10 times before, and puts it back on the rack. He walks away, picks up some batteries that he needs, and returns to the consoles. While there his phone rings; his wife needs him to stop at the grocery store for a few things on the way home. The sensors in the store have already activated the radio chip in the phone and related its identity to the chips on Dave's clothing.

Dave wanders the store, picking up game-playing magazines, looking at CDs and DVDs, each time returning to the game console. But once again he decides against the expensive device, choosing to wait until the price comes down. He pays for the batteries and a magazine with his credit card and leaves. As he pays, the radio sensors at the checkout line register his presence and add two new pieces of data to the store's database: what he bought and the credit card he paid with. The store now knows who he is, how he paid, how much credit he has, where he went while wandering around the store, and a history that shows him coming in again and again to examine the products in the electronic gaming aisle.

The next day, Dave is shocked when he receives a flyer in the mail announcing a sale at the store the following weekend on electronic games. He is equally shocked when he receives a call from the store manager, telling him that he has been selected to attend a private sale event for the store's most loyal customers, during which special prices will apply for invited guests. He is given a code number that will apply to certain products; and while he doesn't know it yet, his number will apply to electronic games and accessories.

Scene-shift now to the forward deployment area of a modern battlefield. Scattered across a wide geography are vehicles, weapons, armament, supplies, medical equipment, and hundreds of troops. Every vehicle has affixed to it a sticky label upon which is attached a small radio tag. Each pallet of supplies is similarly marked, as are weapons, projectiles, and expendable armament. Each warfighter wears a radio button sewn into his or her clothing that uniquely identifies them and creates a logical link to their personnel and medical records.

Overhead, a C-130 cargo plane opens a rear port and spews out a cloud of glitter-like dust that settles to the ground and blankets the area. At the same time the plane ejects dozens of doughnut-sized sensors that fall to the ground and establish wireless connections with the glittery radio devices that now cover the battlefield, creating a de facto wireless internet. The sensors collect and transmit information via secure 802.11 connections from the battlefield array to a local router that analyzes and transmits the data to central command in Pensacola, Florida, where decisions are made based on the received information.

Scene-shift again to the assembly facility of a major computer manufacturer. Over the last year, laptops shipped from the facility have been disappearing from somewhere in the supply chain at an alarming rate, as many as 20 percent of them never reaching their intended destination. To counter the theft and determine where the machines are being stolen, the manufacturer hides these same tiny radio chips inside the laptop cases, while at the same time making quiet arrangements with shipping companies to install sensors along the supply chain so that they can determine where the machines disappear from view, thus pinpointing the point of theft.

In Australia, a veterinarian passes a reader over the paw of a German shepherd that has accompanied its owner from London to Sydney as the result of a job transfer. The reader activates the subcutaneous radio chip that was surgically implanted prior to leaving the United Kingdom, causing it to transmit the animal's medical history to the scanner.[1]

Just down the road, a driver pulls into a petrol station and holds her key chain up to a panel on the pump, which instantly turns on and authorizes the sale of petrol. When she is finished, she sits down behind the wheel and inserts her key into the ignition, which will not start without the proper key because of the presence of a small radio chip embedded in the shank of the key. No radio chip, no ignition.

All of these applications—and many more like them—are made possible by a technology called *Radio Frequency Identification (RFID)*. RFID has been around for quite some time, but is only now becoming a recognized, visible technology. It has been in widespread use in manufacturing and behind the scenes in supply chain applications for years as an alternative to barcode-based product tracking as the products make their way

[1] If the animal had been a cat instead of a dog, the procedure would be a cat scan. (Sorry, couldn't resist.)

from the manufacturer to the consumer. In the last year, however, RFID has gone mainstream, finding application in industries ranging from defense to healthcare, from consumer to enterprise, from supply chain to value chain.

This book is about the practical application of RFID technology. We begin with a brief overview of the technology's fascinating history before diving into how it works in Part Two. We will then spend considerable time in Part Three looking at RFID applications before concluding with a review of RFID products and services and the companies that make them available in Part Four. As always, the book includes an exhaustive acronym list and glossary of terms.

New technologies pop up on a regular basis. Some of them thrive, catching the attention of talented developers with an eye for the market, and become full-fledged products that contribute to the economies in which they find homes. Others flash brilliantly for a few moments like a falling star before lapsing into obscurity. RFID is a technology that is worth watching carefully. It offers tremendous promise in many different industries because it comprises that magical combination of characteristics that often lead to product greatness: the application of an existing, somewhat mundane technology to a set of challenges that lack adequate solutions. RFID is not a killer application; it is, however, a killer technology that elegantly provides a solution to a wide array of business problems. As a result it deserves the careful scrutiny of marketers, application developers, and business professionals, because it isn't going to disappear anytime soon. Quite the contrary: It's here to stay.

As always, I welcome input and comments from my readers. Please write to me at Steve@ShepardComm.com.

Thanks for reading—all the best.

Steven Shepard
Mexico City, Bangkok, Beijing, Williston

PART ONE

BUSINESS ENABLERS

Consider for a moment the staggering chain of events that underlies the movement of mass market products from the producer to the buyer.

In the height of summer in Salinas, California, blood red strawberries are picked, sorted, washed, packed, placed in refrigerated shipping containers, trucked to the airport, and loaded on cargo planes. Nineteen hours later they are on display in supermarkets in Tokyo, Seoul, Helsinki, and Buenos Aires.

Every day, hundreds of thousands of packages pass through the FedEx hub facility at Memphis International Airport. The packages zoom through a high-speed tracking facility at speeds that are almost too fast to watch with the naked eye, moving from one belt to another in a chaotic supply-chain ballet. Ultimately they are fed into trucks, vans, and airplane shipping containers for on-time delivery. Remarkably, there are relatively few human hands involved in the process.

A major manufacturer of personal computers offers custom-built machines with a dizzying array of feature options. Machines are built to order with a promised delivery date, a date that is almost always beaten. To manage costs, the "manufacturer" sources components from a wide variety of suppliers and even outsources the manufacturing process. The only thing they do is provide the direct interface to the customer and manage the overall supply chain. In many ways, they never touch their own physical product.

A major port in the United States sees 200 cargo ships a day come and go, passing through its RO-RO[1] facility as they load and unload thousands of shipping containers, all of which must be registered, weighed, and tracked.

This small list of enterprise activities demonstrates the complexity involved in moving product from a manufacturer or supplier to a buyer. The overall process occurs beneath the aegis of what is known as a *supply chain,* the complex set of activities that result in product delivery management.

SUPPLY CHAINS: A BRIEF OVERVIEW

Supply chains have become a very visible component of the modern corporation. The term is used freely in discussions at the strategic, tactical, and operational levels of corporations—a clear indication of the importance that they have. What they *are,* however, is a very different story. Ask any manager along any part of the product or service delivery process to explain the concept of a supply chain, and, depending on where that person sits in the process, you will hear widely differing responses. For the moment let's retreat to the safety of the dictionary. The *American Heritage Dictionary* defines "supply" as a verb meaning "to make available for use." "Chain," on the other hand, means "a series of closely linked or connected things." A supply chain, therefore, is a series of closely related activities or functions engaged in the process of delivering a product or service, with the intent of making it available for use by some intended recipient. For our purposes a supply chain makes possible the delivery of some physical product or logical service to a customer, often (but not always) via electronic means. For example, when a family moves into a new house and arranges for phone service to be turned on and to include value-added services such as caller ID, distinctive ringing, and three-way calling, the services of a complex supply chain are invoked—yet no physical

[1] Roll-On, Roll-Off: the mechanized technique used in most modern port facilities for loading and unloading container vessels.

product is "delivered" to the customer. On the other hand, when Dell sources components from various suppliers that are installed in the computers that they ultimately deliver to enterprise and residence customers, a similarly complex supply chain ensures that the "mass of components flying in close formation and in the shape of a computer" arrives according to the terms of the agreement between Dell and the customer. In this case, the supply chain ensures the creation and delivery of a physical product.

Of course, this description implies that the supply chain is a one-way path that defines product or service delivery. There are other components of the overall process, however, that must not be overlooked. These components are collectively known as *supply chain management* (SCM). For example, something had to serve as a trigger mechanism that caused the supplier of the product or service to order components and build the computer —an online order placed by a customer, for example. The simple act of a customer going online, designing their machine and then placing the order kicks off a massively complex process that involves payment mechanisms, ordering processes, shipping reservations, billing verification, staffing and scheduling at the assembly facility, and so on. So the supply chain is in fact a two-way process.

When electronic commerce first took off in the 1990s and became a widely used mechanism for selling products into the consumer marketplace, its introduction was not painless. In fact, the first Christmas season brought a number of very large would-be electronic merchants to their proverbial knees, for two reasons. First, their order-acceptance servers were not prepared for the volume of traffic that descended on them. Second, while they did a very good job designing the "order-in, product-out" function, they did not think about how the process for returning merchandise would work! For products purchased at a physical store or through a traditional catalog, merchandise return mechanisms were well established and understood by customers. But when a product ordered online needed to be returned, there was no mechanism in place to handle that process, resulting in retail chaos and a population of disgruntled customers that needed a great deal of attention. Supply chains, therefore, are very much two-way and must be designed with that fact in mind.

Let's take a moment to examine some of the commonly accepted definitions of supply chain management. The *Institute for Supply Chain Management*[2] defines a supply chain as "the design and management of seamless, value-added processes across organizational boundaries to meet the real needs of the end customer." *SearchCIO.com*, a Web site that serves as an information nexus for senior IT managers, defines supply chain management as "the oversight of materials, information, and finances as they move in a process from supplier to manufacturer to wholesaler to retailer to consumer, while coordinating and integrating these flows both within and among companies."

So what do these definitions actually mean? The first definition, from the Institute for Supply Chain Management, speaks to a seamless integration of organizationally independent processes that are designed to meet the ultimate needs of the customer. Each process adds a value component that results in achievement of the end goal, which is successful and accurate product delivery.

The SearchCIO.com definition is similar but incorporates other elements that look backward in the process, as well as forward. This definition speaks to the overall process management that involves oversight of acquisition of raw materials, incorporation of information, and judicious use of financial resources to ensure product delivery.

Jack Garrett is a global thought leader in supply chain management who specializes in designing the intricate processes that make these complex systems work. He offers this definition: "The Supply Chain is the set of policies, processes, management actions, and technologies that collectively forecast, acquire, and deliver products and services to meet the identified needs of a company for its own internal purposes and for its delivery of products and/or services to its customers." Note that in this definition, there are multiple defined layers that the supply chain affects. There is the provider of the purchased service

[2] For more information go to www.napm.com.

or product; the purchaser of that same service or product; and the ultimate beneficiary of the service or product, the so-called customer's customer. "The supply chain extends throughout and defines the end-to-end relationship that exists between the buyer and the seller, as well as among all the providers of ancillary functionality.

"The overall supply chain is complex, and the management process is equally complex," says Garrett, "referring to the diagram shown in Figure 1-1. "There are actually three phases to the overall process—Procurement, Purchasing, and Support.

"The diagram illustrates the overall flow pretty well. Procurement begins with customer business requirements, the creation of a statement of work and RFx,[3] vendor selection, contract negotiation, and so on. Purchasing involves the creation of a purchase order, the shipment of the finished goods, receipt of the finished goods, and the billing and payment

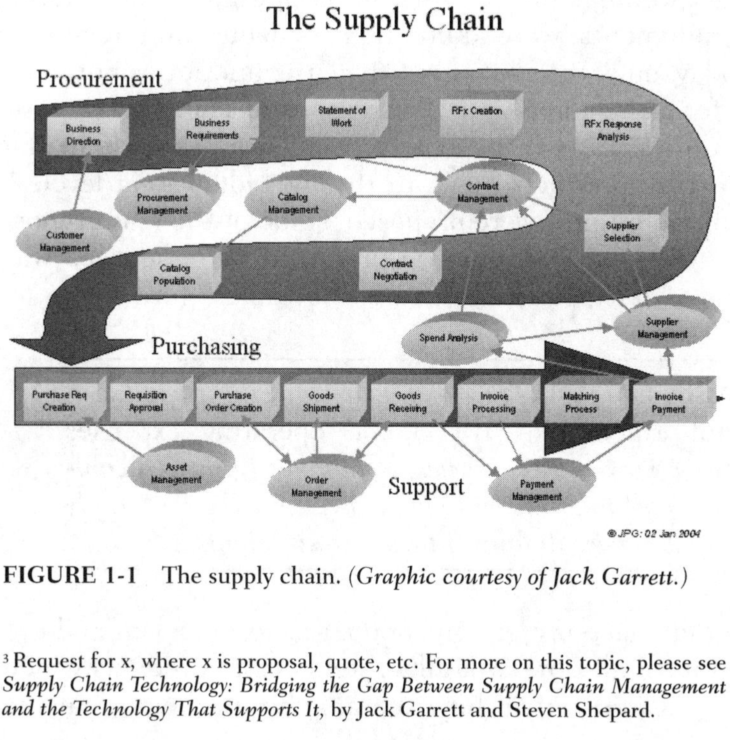

FIGURE 1-1 The supply chain. *(Graphic courtesy of Jack Garrett.)*

[3] Request for x, where x is proposal, quote, etc. For more on this topic, please see *Supply Chain Technology: Bridging the Gap Between Supply Chain Management and the Technology That Supports It*, by Jack Garrett and Steven Shepard.

processes. Support occurs throughout the supply chain, with intelligent management processes applied to accelerate and tighten the overall mechanism of the supply chain."

As products move through the complex maze of the supply chain process described here, they must be tracked at every step of the process. Such tracking activities help vendors control costs, streamline the delivery process, reduce theft, identify weaknesses in customer relationships, strengthen relationships with other functions and groups involved in supply chain management, tighten spending control, and better understand asset management, which is an enormously expensive function. As the number of managed items increases, as the size of the market swells from local to regional to national to international, as payment and ordering options evolve, and as the market and the products it buys become more and more complex, the ability to track and manage individual items throughout the end-to-end supply chain becomes unwieldy and difficult. When the market was small and simple, so was management. It was quite enough to manage products at the truck or shipping container level, because shipments were quite homogeneous and relatively small. Today, however, that is not the case, and it has not been the case for quite some time. For reasons of competitive positioning and business management, a need arose early in the 1930s to track products down to the individual item level. A number of techniques were proposed, some of them as preposterous as they were unique. The first of these to succeed was the barcode, an example of which is shown in Figure 1-2.

In reality, companies are looking for four things: reduced capital expenses (CAPEX) *and* operating expenses (OPEX)*; increased revenues; a stable or improved competitive position; and mitigated risk. An effective supply chain addresses all four of these requirements.*

Supply chains represent one of the greatest chicken-and-egg conundrums of all time. Manufacturers create products which they hope customers will buy; the assumption being that the

FIGURE 1-2 A typical barcode, this one for a book.

customer has either actively committed to a desire for the product or an influence by the manufacturer has led them to believe that demand will grow (the Kevin Costner Effect—"If you build it, they will come"). Of course, to build the product, new materials and components were purchased from upstream suppliers. Those suppliers had to gear up and manufacture as well, with production based on decisions similar to those discussed earlier. Of course, there are other functional layers involved. An agreement layer exists, where contracts are created that guide the relationship between the buyer and the supplier; a payment-for-services layer exists, where invoices are generated and payments are returned; a product delivery assurance layer exists, which defines service or product quality and the consequences for failure to deliver under the terms defined by the contract layer; and so on. As a product[4] traverses the gap between the supplier and the buyer, it passes through a long and complex series of processes known as the supply chain. The supply chain is the collection of activities that must occur in a carefully orchestrated and synchronized way to ensure that the right product reaches the right customer at the right time for the right price, and that value is enhanced at each stage along the way can be monitored to ensure that value is enhanced at each stage. This is reminiscent of the fishbone diagrams (Figure 1-3) that have been described in

[4] The term *product* is used generically; it could just as easily be a paid-for service.

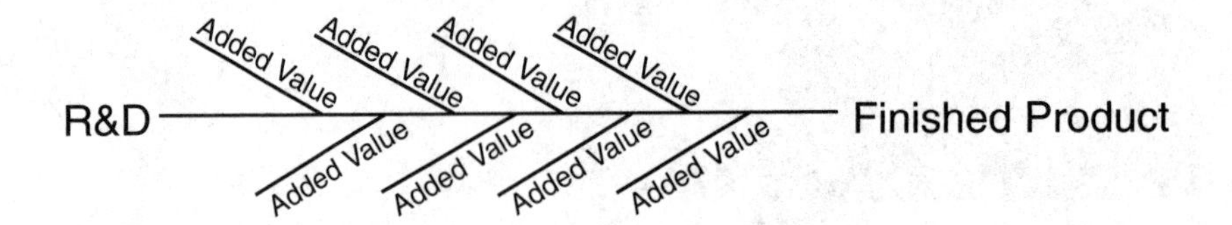

FIGURE 1-3 A fishbone diagram: At each stage of the process, value is added to the overall outcome in the original concept that led up to the "value chain" theory.

business books for years, where each "rib" of the fish represents the addition of some component of value along the way to product delivery.

In reality, the supply chain is a value chain. As products move through it, they must be tracked to ensure that the stipulations of the contractual agreement are met along the way. Value is a term that is thrown around routinely, often without much thought behind its use. How do supply chains deliver on the promise of added value? The truth is that they deliver it in four ways. Coincidentally, they are the same four things that companies list most commonly as their critical success drivers: reduction of CAPEX and OPEX; increased revenues; stable or improved competitive position; and mitigation of risk. And how does an effective supply chain contribute to these rather lofty goals? To answer that question, let's look at a common (albeit hypothetical) example of a company whose supply chain contributes measurably to the company's success.

THE SERVICEPLUS STORY

ServicePlus is a membership-based retail store chain that targets *small-to-medium-sized businesses* (SMB) as its primary customer segment. This includes the residence market since a significant percentage of small businesses are operated out of homes. The company has more than 2,000 warehouse stores in North America alone and a thriving Web-based business. They sell office supplies, office furniture, and computers, which they build on demand to customer specifications. Furthermore, about 25 percent of each store's floor space is devoted to "ancillary products"—not technically office supplies, but products that are often purchased by businesses, such as dry goods and perishable foods, drinks, cleaning products, party supplies, paper products, small appliances, and telephone and data products. The company has its own brand for many *original equipment manufacturer* (OEM) products, but also sells products from more than 1,000 different vendors.

In addition to the 2,000+ stores in North America, ServicePlus also has warehouse stores in Europe, the Middle East and Africa (EMEA), Asia-Pacific (Singapore), and China (Shanghai and Hong Kong).

This is a company with *enormous* purchasing power, which translates into equally enormous influence over its suppliers. ServicePlus aggressively negotiates prices with its suppliers, but at the same time, the company offers unequalled sales opportunities to those suppliers through its role as a retail portal.

When ServicePlus came into existence in the early 1970s, it started as a small cluster of stores in California with a headquarters facility in Rohnert Park. Its IT department, rudimentary though it was, was housed in the headquarters data center, and as the company grew it became more complex in keeping with the founder's fundamental vision that centralized information management would help the firm manage itself and its relationships with customers more effectively. Every night the original four stores would upload their sales (and therefore inventory) data via dial-up line to the mainframe-based management application in Rohnert Park, and from that application were generated sales figures, inventory reports, and shipping lists on a store-by-store basis.

As the company grew and stores were added to the system, it became clear that dial-up was far too slow, cumbersome, and expensive. In response, the IT organization entered into an agreement with a *very small aperture terminal* (VSAT) provider, allowing each store to uplink to a satellite and upload data via the faster wireless connection. More recently the VSAT network was replaced with a secure Internet-based *virtual private network* (VPN).

Inventory is monitored in each store in two ways. When a customer checks out and pays for his or her purchase, the laser scanner reads the barcode on each item, causing the customer's invoice to be incremented and the store's inventory to be decremented. Secondarily, ad hoc inventory of items that are prone to shrinkage (theft) is routinely conducted using handheld readers that send information back to the store server for reconciliation and verification. This is done via a secure Wi-Fi connection.

ServicePlus has a very tight and strictly controlled relationship with its suppliers. Because it monitors inventory so carefully and because it holds its suppliers to such high standards of product quality, availability, and inventory control, the firm's IT organization has created logical interfaces between its own internal purchasing systems and those of its suppliers so that they can have limited access to the ServicePlus database. Because suppliers now have access to product depletion-rate data on a real-time, store-by-store basis, they can (and do) monitor the inventory levels of their own products. In fact, thanks largely to work done by ServicePlus's IT professionals, the entire inventory monitoring and management process is mechanized. As products pass through the barcode readers at checkout, or are scanned by handheld devices, ServicePlus's databases are updated in real time. Those databases are monitored by supplier IT management systems. Based on known purchase volume histories in each store, when levels of each product reach a low watermark, the supply chain is automatically triggered, generating a restock order to the supplier for that particular product in that particular store. As a result of this just-in-time relationship between the ServicePlus system and suppliers' systems, stores rarely run out of products before they are restocked, resulting in (1) higher customer satisfaction levels, (2) lower inventory costs, and (3) avoidance of aging inventory. At the same time, automated processes from the supplier generate an electronic invoice to the ServicePlus accounts payable system, which in turn triggers an automated payment response that occurs according to the parameters of the contract that is in place between the two, part of which dictates payment terms.

The combination of supply chain mechanization and the buying power of such a large retail company goes well beyond the influence that ServicePlus has over its suppliers. The firm's influence also extends to the shipping companies they rely on for in-store stock resupply and deliveries that result from Web-based sales. When a store restocking order, custom computer order, or Web-based sale result in a shipping request, automated interfaces between ServicePlus and the shipper instantly link the shipment to a tracking number so that the store and the customer can track the progress of product delivery throughout the end-to-end process.

There are numerous other parts and components of the supply chain mechanism described here, including RFx generation, catalog management, inventory rotation, and warranty enforcement. But even taking into account this small subset of the processes involved in moving products from a supplier to a buyer, the reader should have a sense of the enormous complexity of the process and the degree to which supply chain mechanization hides that complexity from the user of the system.

Think now about how the process described in the last few paragraphs affects the four critical success factors listed earlier. OPEX and CAPEX are reduced in a variety of ways. If an enhanced supply chain results in the reduction of on-hand inventory in favor of just-in-time practices, then OPEX is reduced as a result. Additionally, supply chain mechanization means less humans are involved in the supply chain process, which further reduces OPEX. CAPEX, on the other hand, is reduced because of the ability to reduce warehouse space and product handling equipment.

Revenues are increased as a result of supply chain management for one simple reason: customer loyalty. When an effective supply chain management program translates into the right product delivered to the right customer in the right place for the right price at the right time, customers remember—and come back.

Enhanced *competitive advantage* results when a business operates more cost effectively than its competitors, or when it consistently offers a level of service that is measurably better than the competition. Effective supply chain management is designed to do precisely that.

Finally, a well-designed supply chain can dramatically *mitigate risk* for a business by reducing on-hand inventory through just-in-time monitoring processes, by accelerating the accounts payable process, and by aggressively managing accounts receivables.

We turn our attention now to the facilitative technologies developed early in the supply chain management field, beginning with barcodes.

THE HISTORY OF BARCODES

Early in the twentieth century, the most complex business in existence (from a supply and demand and the inventory-on-hand perspective) was unquestionably the grocery store. Grocery managers had to stock tens of thousands of items in various sizes from a wide variety of suppliers, manage the perishable items differently than the nonperishable stock, and ensure proper rotation of on-hand inventory, all while operating in a business environment that offered extremely small profits. In the early days, stock-on-hand was calculated by physical inventory: shutting down the store and setting all employees to the task of counting items on the shelf. Because there were humans in the loop, this process was slow, tedious, expensive, and error prone. As a result, many store managers simply estimated their inventories, resulting in numbers that were a far cry from accurate. It was clear to grocery managers that they needed some form of mechanized tracking, and they needed it fast. Oddly enough, the first solution to their problem came as the result of the U.S. census.

THE 1790 CENSUS

In 1790, it took the Census Bureau about eight months to complete the first U.S. census. By 1860, however, the population had increased almost tenfold, from 3.8 million to 31.8 million. In 1880, the Census Bureau initiated another national census; in 1887 they completed it—*a full seven years later.* The size of the population, the immense geography over which it was spreading, and the primitive nature of the technology that was available to help with the collection and tabulation of data made it virtually impossible for census takers to complete their task within a reasonable timeframe.

The need for an accurate population count was not purely scientific: It was required to meet the mandates of the Constitution, under which representation of the people is calculated upon population. The 11th census occurred during a time of unprecedented growth in the United States, a time when populations were growing extremely diverse and spreading throughout the country in pursuit of their own piece of the Manifest Destiny. Between 1880 and 1890 the national population increased by more than 12 million people, and many of them, immigrants seeking freedom and prosperity, struck out for the western territories, which further complicated the job of the Census Bureau. As a result, the data collected during the 11th census was out of date before the census was completed!

In a rare spark of government ingenuity, the Census Bureau decided to hold a contest. The winner of the contest would

FIGURE 1-4 Herman Hollerith, inventor of the machine that mechanized the Census Bureau's data collection processes. *(Photo courtesy Rochester University.)*

submit the most ingenious method for mechanizing the census process. As it turned out, the winner was Herman Hollerith (see Figure 1-4), a native of Upstate New York and a recent graduate of the Columbia School of Mines, who later went on to become an employee of the Census Bureau for a time. Hollerith designed and built a machine that not only mechanized the data collection and tabulation process; it saved the Census Bureau more than $5 million.

Hollerith's machine, shown in Figure 1-5, was based on an earlier device created by Jacques Jacquard, an engineer who was active during in the Industrial Revolution. Jacquard specialized in weaving technology, and somewhere along the way he recognized that the process of weaving cloth is a highly repetitive task that, while complex, could be mechanized to speed up the process and improve the accuracy of the weaving. He designed a machine that relied on rigid cards with patterns of punched holes in them to automate the weaving process. At each throw (forward movement) of the loom's shuttle, a card was placed in the path of the pattern rods. The pattern of holes in the card determined which rods could pass through it, resulting in a set of instructions that the loom was compelled to follow—in effect,

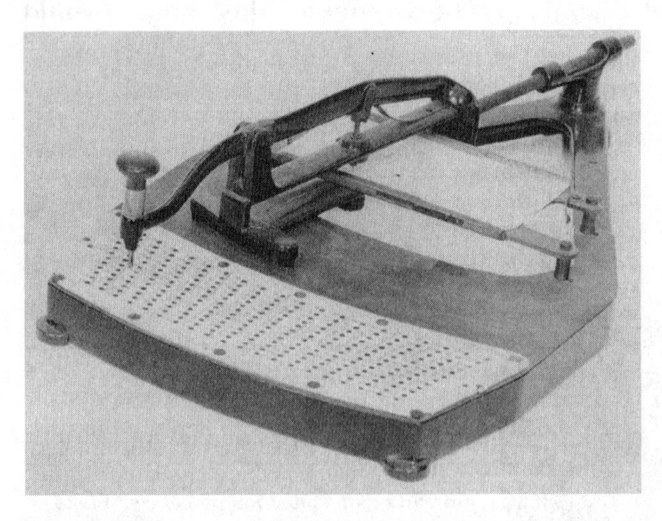

FIGURE 1-5 The Hollerith Pantograph.

the first machine "program." And while the technique was designed to mechanize the process of weaving cloth, it was identical to the process that would later punch cards for computer data entry.

Hollerith was familiar with Jacquard's work and felt that it could serve as a basis for his own mechanized counting system; in fact, his first attempt was based on Jacquard's machine. Hollerith used the punched card concept in concert with a continuous paper feed. The holes, arranged in two rows as shown in Figure 1-6, represented the items to be tabulated; male vs. female, immigrant vs. native born, and so on. A later version used combinations of holes to represent more complex information such as geographical data. In 1884 he described his method in a patent application: "Various statistical items for a given person are recorded by punching suitable holes in a line across the strip, being guided by letters on the guide plate." The original patent is shown in Figure 1-7.

The machine was deceptively simple. The paper was punched by census takers as they collected and input data during their daily activities. The strip of paper was fed into the counting machine, where it passed over a drum with an electrified surface.

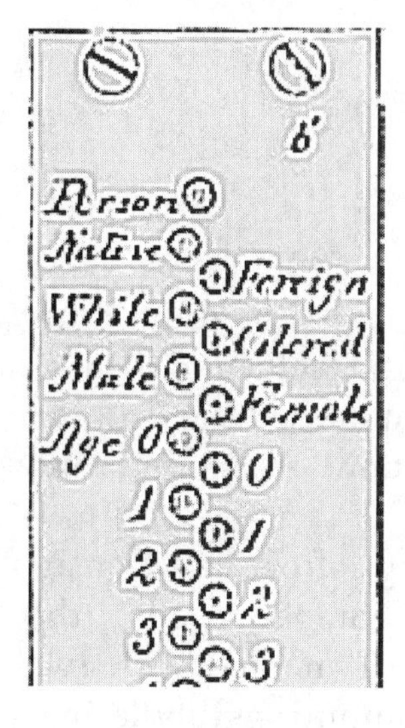

FIGURE 1-6 An example of the card used in the Hollerith Pantograph for tabulating census data.

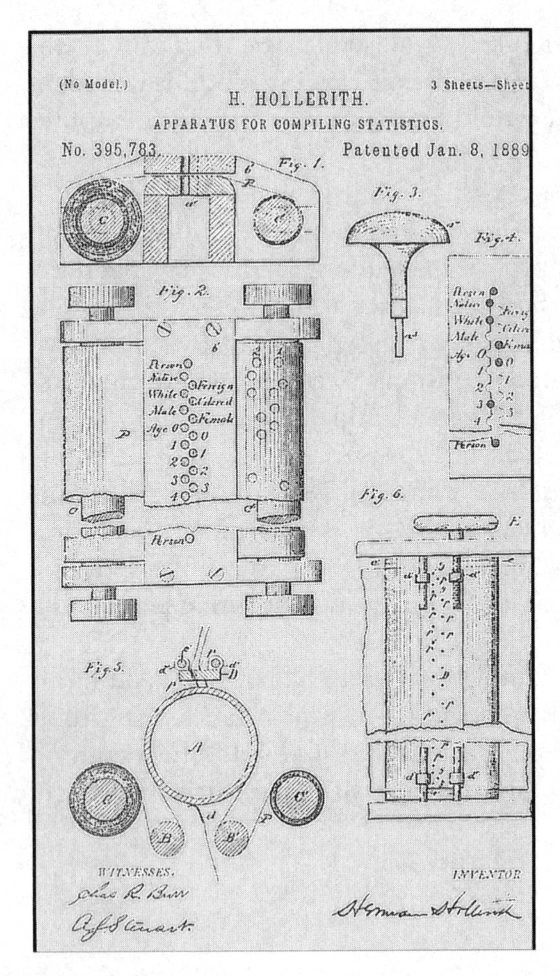

FIGURE 1-7 Hollerith's original patent.

If a hole appeared in the strip as it passed across the drum's surface, a circuit was completed between a metal feeler and the surface of the drum. Completion of the circuit caused the counter, corresponding to that piece of information, to register and increase the count of that particular item.

Hollerith's successes continued long after his Census Bureau career. In 1890 he founded a company called the Tabulating Machine Company. In 1911 it merged with two other companies to create the Computing-Tabulating-

Recording Company. Under the management of Thomas Watson, Sr., the Computing-Tabulating-Recording Company would change its name in 1924 to International Business Machines. Hollerith's punch provided the foundation for IBM's success, and made him the father of modern computing.

From Census . . . to Groceries

In 1932 Wallace Flint, a business student interested in the field that would one day become Business Process Re-engineering, wrote a master's thesis in which he described an inventory-intensive retail business, such as a supermarket, where customers would select their groceries by punching holes in cards similar to those designed by Jacquard and adapted by Hollerith. Once their cards were punched, they would take them to the checkout counter where they would insert their punched cards into a reader and pay for their selections. Meanwhile, the reader would activate a sequence of events that would result in each customer's groceries coming to them on a conveyor belt. As a result of this mechanization process, store managers would have an accurate record of inventory-on-hand as well as a record of purchases on a customer-by-customer basis, giving them the ability to analyze customer buying habits.

Flint's idea was a good one, but it was fraught with challenges. The country was in the stranglehold of the Great Depression, and few, if any, small businesses could afford to purchase something so outlandishly expensive and complex. The idea was ahead of its time and was not implemented. It did, however, form the basis for a series of innovations that would lead to the development of barcodes and modern supply chain management, as we know it today.

In 1948 graduate student Bernard Silver was loitering in the halls of Drexel Institute of Technology in Philadelphia. Standing there, waiting for a class to begin, he listened to a conversation going on nearby between the president of a large grocery store chain and one of the institute's academic deans. The grocery executive wanted the dean to commission a research project into the feasibility of designing a system that would automatically record product information as items were purchased during the checkout process. The dean declined the request, but Silver pondered the request for several days, finally describing the conversation to Norman Woodland, a 27-year-old graduate student and teacher at Drexel. The problem intrigued Woodland, and he agreed to consider it as a possible project.

After much deliberation, Woodland came up with a number of possible techniques for capturing and recording product information during the purchase process. There were clearly two components required to make the process work. First, each purchased item had to have a unique identifier; second, there had to be a device at the checkout counter that could read and capture the information printed on each product.

Woodland's first idea was to use patterns of ink that would fluoresce when placed under ultraviolet light. Woodland and Silver built a prototype to test the concept, and it worked, but other problems presented themselves, including the cost of printing identifiers on every product and the relative instability of the inks that were available at the time.

Woodland persevered, however, and became convinced that his concept was sound and would prove to be valuable to the retail world. Stealing himself, he cashed in some stocks that he earned, quit his position at Drexel, and moved to Florida, where he set up a lab in his grandfather's house. He worked steadily for many months, after which he emerged with the first of several innovations: the linear barcode.

Woodland's barcode was different from those seen commonly on products today. His design relied on the combination of two concepts taken from different, disparate industries: the encoding scheme used in movie soundtracks, and the traditional dots and dashes of Morse code.

In the early 1920s, when silent movies were all the rage in the United States, inventor Lee DeForest created a technique for adding sound to the movies, resulting in the first "talkies." His technique involved reserving a strip along the edge of the film and striping it with a pattern that had varying levels of opacity. A light shining through the stripe on the moving film struck a photosensitive tube on the other side, which converted the varying levels of light intensity into varying levels of sound. Woodland's intent was to use this same technique, but rather than convert the light signal into sound, his system would convert it into recordable data. The pattern he used was based on the dot and dash pattern of Morse code; he simply extended the dots and dashes in a linear fashion, "smearing" them into long patterns of thick and thin lines.

Woodland's encoding scheme was relatively straightforward. It comprised a simple pattern of four white lines on a dark background. The first line was a demarcation line, and the positions of the remaining three lines were fixed with respect to that first line. Information was encoded as a function of the presence or absence of one or more of the three lines. With three lines, six product classifications could be uniquely identified. It didn't take Woodland and Silver long to recognize that by adding lines to the pattern they could identify significantly more products; with ten lines, more than 1,000 items could be uniquely encoded.

In the lab, Woodland's idea worked; he decided to take it to the next level.

Returning to Drexel, Woodland began the long and convoluted process of patent design. Along the way, he decided that the linear barcode idea did not work as well as he had anticipated because the pattern had to be scanned from a particular direction. Failure to do so resulted in gibberish. He replaced the linear design with a pattern of thick and thin concentric circles that could be scanned from any direction and that came to be known as the *bull's-eye code*. An example of the bull's-eye code is shown in Figure 1-8. On October 20, 1949, Woodland and Silver filed a patent for their design.

Shortly thereafter, in 1951, Woodland applied for and was given a position at IBM, where he intended to pursue his idea of a digital encoding scheme for product identification. In 1952 he and Silver built a prototype barcode reader in Binghamton, New York, and while the device worked, it was huge. It was roughly the size of a washer and dryer, and because the photo detector that resided inside was so sensitive, they had to wrap

FIGURE 1-8 The bull's-eye code, shown here in the upper-left corner.

the entire thing in thick black oilcloth to keep out ambient light.

The reader consisted of two key elements: a 500-watt (!) incandescent bulb which served as the light source, and an RCA 935 photo-multiplier tube, originally designed for movie sound systems, which served as the receiver.

To test the efficacy of the prototype, Woodland attached the RCA 935 photo-multiplier to an oscilloscope. He then positioned a piece of paper printed with a barcode pattern between the light source and the photo-multiplier tube, and moved it slowly across the space between the two devices. As the paper moved, the displayed signal on the oscilloscope changed, indicating that the device worked. Never mind that the heat from the 500-watt bulb set the paper on fire (no doubt a cause for concern in grocery stores). The device functioned properly: Woodland and Silver had created a system that would read information electronically.

Of course, it wouldn't do for grocery checkout clerks to be required to sit before an oscilloscope screen and interpret the sinusoidal waveform representing a can of soup. The information represented by the oscilloscope display had to be converted into a form that was actually useful and that could be captured. Keep in mind that we're talking about the early 1950s: Computers of the time were measured in acres and cycles-per-week. The idea of deploying them in grocery stores across the country was—well, silly. But the challenge remained: Without a capture and conversion mechanism, the barcode concept would never achieve commercialization.

One of the biggest challenges that the two inventors faced was the heat put out by the 500-watt light source. The bulb generated an enormous amount of light (not to mention heat), which meant that the device was extremely inefficient and dangerous. Unfortunately, there was no easy solution available. Lasers were still years away from commercial development, and the concept of a light-emitting diode that would operate more efficiently (with less heat generation) hadn't even been dreamed of yet.

In October of 1952, after a few years of wrangling with the Patent Office, Woodland's and Silver's patent was granted. Woodland remained with IBM, continuing to work on the concept of an electronic reader, and in the late 1950s persuaded the company to hire a consultant to evaluate barcode solutions. After evaluating the market and assessing the viability of a barcode reader in industry, the consultant agreed that the device was viable—indeed, desirable—but that it would require technological innovations that were still quite a few years away.

IBM, always good at spotting a good thing with future potential, offered to buy the patent from the two inventors on several occasions, but never for a price that Woodland and Silver felt was adequate. Finally, in 1962, Philco came forward with a reasonable offer and they decided to accept the offer. Philco later sold the patent to RCA.

So let's take stock. Woodland and Silver have invented a device that will electronically read data from labels printed on commercial products. Unfortunately, other than for prototyping purposes, the technology does not yet exist to make the idea commercially viable. Furthermore, a keystone application has not yet been identified that will help to create critical mass for the concept. One did emerge, however, and its arrival on the scene provided the necessary impetus to the barcode reader concept. That application was the need to track freight cars on the nation's railroad system.

TRACKING THE RAILROADS

While every freight car that crisscrosses the nation's railroad system is owned by a particular railroad (the name is typically emblazoned on the side of the car), the cars are actually treated as a massive lending library, constantly being loaned back and forth from one freight company to another. Nevertheless, they need to be tracked, if for no other reason than to assess rail miles for maintenance scheduling. The process of maintaining an accurate inventory of rolling freight is massively complex, and it was no less complex in the 1950s when David J. Collins was involved in the task at the Pennsylvania Railroad, where he worked while attending MIT as an undergraduate.

When he graduated in 1959, Collins went to work for the Sylvania Corporation which, at the time, was consumed with the dream of finding military applications for a computer it had recently designed and built. Collins knew that railroads needed a way to automatically identify rolling stock and to collect and interpret the information gathered. He also realized that the computer that Sylvania was struggling to find an application for was ideally suited to the task of tracking railroad cars. Of course, the original problem still remained: how to uniquely identify each car so that a reader could collect the identity data and feed it into the Sylvania computer? Clearly a coded label, along the lines of what Woodland and Silver had conceived, was in order. Adapting the ideas conceived by Woodland and Silver, Collins developed a technique that relied on groups of orange and blue stripes made of reflective material that could be arranged on the sides of rail cars to represent the digits zero through nine. Each train car was assigned a four-digit number to identify the railroad that owned it, and a six-digit number to identify the car. When cars rolled into a train yard, a reader flashed a beam of colored light onto the identifying pattern on each car, and interpreted the reflected light. In 1961, the Boston & Maine Railroad agreed to conduct the first test of the system on its gravel cars. It worked; by 1967 most of the issues had been worked out of the system and a nationwide standard for car coding was adopted.

Collins, meanwhile, had already moved beyond the railroad industry. He watched technological advances closely and knew that it was only a matter of time before computers would be cost effective enough to be installed in smaller industries than the railroad. And based on his foresight, he went to the senior management of Sylvania and made the case for further development of the electronic reader concept. His goal was to create a barcode model that would work in factory assembly line environments and that could be applied to process control. Unfortunately for them, Sylvania declined to fund the effort, so Collins left. With a partner, he cofounded Computer Identics Corporation, dedicated to creating practical applications for the technology that Collins had developed based on the earlier work of Woodland and Silver.

In 1970 railroads began the task of installing scanners throughout the railroad system. It worked precisely as it was intended, but unfortunately, it proved to be more expensive than it was worth at the time. Even though computers had continued to grow smaller and less costly, their implementation was still not adequately cost effective for the numbers that needed to be installed. When railroads began to declare bankruptcy during the 1970s recession, the project withered and died.

In the meantime, Collins's Computer Identics prospered. Its system relied on lasers, which by the late 1960s had just become affordable for such applications. Collins's new design relied on a thin light beam that moved across a barcode, which reflected the light in varying levels of intensity. Because of their coherent nature, lasers could read barcode data over a significant distance range (three inches to several feet). Furthermore, they could be designed to scan a code from many different angles searching for the proper orientation, which overcame the problem of product orientation on the conveyor belt as well as the challenge of torn or smeared labels.

In 1969 Computer Identics installed two systems. One found its way into a General Motors plant in Pontiac, Michigan, where it monitored the production and distribution of axles. The other was purchased by the General Trading Company in Carlsbad, New Jersey, where it was installed in the firm's distribution facility and used to direct shipments to the appropriate loading bays. These were the first successful implementations of commercial barcode readers. The coding scheme they used was a simple two-digit configuration that could encode as many as 100 unique identifiers (0 to 99), but since the Pontiac plant made only 18 different axle types, a two-figure code was more than adequate for their needs.

BACK TO GROCERIES

Keep in mind that RCA owned the patent for Woodland's and Silver's original design, and while they had been notably absent from the public eye during Collins's railroad adventures, they were still in the game. In 1966 RCA executives attended a grocery industry conference where a major item of discussion was the use of barcodes for product identification. Intrigued by the potential of a new and potentially lucrative line of business, they sent a study group to an RCA laboratory in Princeton, New Jersey, and at the same time arranged for the Kroger grocery chain to serve as a volunteer in their efforts to develop a commercially viable system for the grocery industry.

While this was going on, other events were taking shape as well. In 1970 a consortium of industry representatives formed a committee to examine the feasibility of barcode implementation. They established development guidelines and formed a symbol selection committee with the task of standardizing the data representation methodology.

The committee's work was quite successful for one key reason: Members approached their task from the perspective of the audience that would have to use the technology rather than from that of the technologists developing it. Their guidelines, for example, were simple. First, barcodes would have to be readable from any angle and at a range of distances. Second, barcode labels would have to be cheap and easy to print, because they would have to appear on every possible product if the concept was to succeed. Finally, the implementation would have to be cost effective, with a 30-month payback period. This turned out to be eminently doable; a feasibility study conducted in 1970 by McKinley & Company demonstrated that the industry would save more than $150 million if it adopted the barcode model. The challenge, of course, was getting management to buy into the idea and commit to universal barcode labeling.

In early 1971, during a grocery industry conference, RCA demonstrated a bull's-eye barcode in a way that made a profound impact on the attendees of the conference. Each attendee was given a small round piece of metal on which was printed a code. They were encouraged to pass their code beneath a scanner, and if they held the right code, they won a prize. Needless to say, convention goers were no different then than they are today; the RCA booth buzzed like a beehive.

This did not go unnoticed by other exhibitors. IBM executives a few booths away noticed the crowds and began to wonder what they were missing out on in the emerging retail marketplace. At about the same time, IBM marketing specialist Alec Jabionover remembered that the barcode's inventor (remember Woodland?) worked for IBM! Even though his patent had expired in 1969, Woodland was quickly moved to IBM's research and development facility in North Carolina, where he would ultimately play a key role in the development of the *Universal Product Code* (UPC), which became the most widely distributed and successful version of the barcode concept and is still in use today.

Building on the seeming success of its bull's-eye code and the overwhelmingly positive response it received at the grocer's conference, RCA continued to drive forward with its commercial development. In mid-1972 it began an 18-month trial of the technology in a Kroger store in Cincinnati, Ohio. Unfortunately, printing problems (cost of ink and smearing) and accurate scanning issues prevented the bull's-eye model from achieving commercial success. Ultimately, IBM's UPC model was chosen by the industry. Adopted on April 3, 1973, the UPC revolutionized inventory tracking and management in the retail world.

The UPC is divided into two six-digit components. The first digit is always zero except for meat, produce, and other products that have variable weight. The next five digits represent the unique manufacturer's code, while the following five digits are the product code. The final digit is used as a checksum to ensure that the prior eleven digits have been successfully (and accurately) scanned. Structural variations in the barcode indicate the orientation of the barcode to the scanner, allowing the product to be scanned from any direction.

Technologically, it was the sudden availability of the laser and the integrated circuit that made the barcode scanner feasible. As these technologies became mainstream and found their way into more and more facets of life, they became accepted as routine and helped propel the barcode model toward critical mass and widespread acceptance. On June 26, 1974, at a supermarket in Troy, Ohio, a barcoded pack of chewing gum became the first retail product sold with a scanner. The technology penetrated the market slowly; IBM and others knew that 85 percent or more of all products would have to be imprinted with barcodes if the system was to be successful, and in the late 1970s, they surpassed that figure. Still, it wasn't enough to have a barcode—stores had to be able to read them. Less than one percent of grocery stores had barcode scanners in 1970; by 1981 that number had climbed to 10 percent, and today the number is well over 85 percent.

BARCODES . . . UP CLOSE AND PERSONAL

The UPC was the first broadly accepted barcode model. Originally adopted by the grocery industry, it quickly became the most widely used standard for product identification. It succeeded so well in the marketplace that international interest developed, resulting in the creation of the *European Article Numbering* (EAN) system and the *Japanese Article Numbering* (JAN) system.

Today there are five versions of UPC identification and two versions of EAN. The JAN system, discussed later, is identical to one of the EAN versions in which the flag character is set to 49, uniquely identifying it as a product identification scheme for the Japanese market.

Under the current rollout plan, all U.S. retailers must be able to scan 8-, 12-, 13-, and 14-digit EAN/UPC codes by 2005.

ENCODING DETAILS

UPC and EAN symbols are fixed-length numeric-only entities. Examples of the most common are shown in Figure 1-9.

UPC Version A comprises ten identification digits plus two overhead digits, while EAN symbols have 12 identification digits and a single overhead digit. The first overhead digit of a UPC Version A identifier relates to the type of product, while the first two digits in an EAN symbol designate the country of the EAN international organization that issued the number. And because UPC is a functional subset of the EAN code, scanners that read EAN symbols can also read UPC symbols. UPC scanners, however, cannot necessarily read EAN symbols—hence the 2005 mandate.

One of the key challenges that faced the creators of barcode systems was the actual process of scanning and reading the

FIGURE 1-9 The most common forms of barcode: UPC Version A is shown in the upper left; EAN-13, upper right; EAN-8, lower left; and UPC Version E, lower right.

information contained on the marked package. The UPC coding scheme was designed to overcome one of the major issues—product orientation. Thanks to significant forethought, the format of the UPC allows it to be scanned from any angle, making the orientation of the product immaterial. According to studies, UPC Version A has a first-pass read rate of 99 percent using a fixed laser scanner, and a substitution error rate of less than one error in 10,000 scans.

UPC Version A is typically found on grocery store products and is most commonly used to encode the 10-digit UPC, along with an 11th type-of-product identifier digit and a 12th data-check digit (used for error detection). The Version A symbol is divided into two six-digit halves. The six-digit fields are surrounded by left, center, and right guard patterns. The left field is encoded using odd parity; the right uses even parity. As was described previously, the first digit identifies the type of product, while the following five digits identify the UPC manufacturer's code. The first five digits of the right half represent the product code, while the final digit is the checksum. And while UPC A is considered to be a continuously encoded field, the left and right halves of the symbol can be encoded independently.

DIGIT ENCODING IN UPC A

A digit is encoded as a sequence of two bars and two spaces within a space that is seven modules wide (see Figure 1-10). The bar and space widths can be 3, 4, or 12 modules wide, resulting in 20 possible bar-space combinations. Ten are used for the *left* odd parity digits, while the other 10 are used for the *right* even parity digits. The left digits always begin with a space, while the right digits always begin with a bar.

A typical Version A symbol has guard bars in the center of the symbol which are longer than the other bars. These guard bars divide the symbol into a right and left half, which provides the means for a scanner to read the symbol from any orientation. The laser-based barcode scanner generates orthogonal scanning beams in a variety of patterns, including figure eight, cross, or starburst. Because of the plethora of beams that flash

FIGURE 1-10 The structure of a common barcode layout.

across the scanned product, at least one will pass through each half of the symbol.

Ideally, the height of the printed symbol should be at least half the length of the symbol, although this is often violated in deference to product packaging requirements. This can affect the ability to orthogonally scan and can reduce the first-pass read rate.

In addition, the so-called quiet zZone (that is, the area on either side of the UPC that has no printing) should be nine modules wide on the left and right sides of the symbol. In some cases, Version A codes may contain an additional two- or five-digit supplement, often used on books or magazines, as shown in Figure 1-11.

UPC VERSION E

The second most common version of UPC codes is Version E, intended for use on packaging that is physically too small for the other versions. It is known as a *zero suppression version* of UPC, because to achieve the smaller size it drops the zeros that would otherwise occur in the coded identifier. For example, the code shown in Figure 1-10, which is for a package of guitar strings I recently purchased, would be encoded as 29789112

FIGURE 1-11 A barcode symbol created to display the ISBN information of a book.

instead of 2978910120. The final digit indicates the type of compression that was used to suppress the zeroes. Guard bars precede and follow the data, and the middle guard bars are eliminated. The digits are coded according to an even, even, odd, odd, even, odd parity scheme, and the data is encapsulated between two left-hand guard bars and three right-hand guard bars. The resulting six-digit number is always preceded by a zero and followed by the check digit.

OTHER VERSIONS

There are three other versions of UPC which are far less commonly utilized for product identification and tracking. Version B was originally developed to identify products that fell under the National Drug Code and National Health Related Items Code. It accommodates 11 digits plus a product-type code, but does not include a checksum.

Version C was somewhat heroically created to promote industrywide compatibility. The code is 12 digits long and includes both a product-type digit and a checksum digit.

Finally, UPC Version D is a variable-message-length UPC. It must contain at least 12 digits, the first of which is a product-type code. Next is a 10-digit information field, followed by a checksum in the 12th position. The checksum can then be followed by a variable number of digits.

EAN-13 AND EAN-8

The EAN, the JAN, and the *International Article Numbering System* (IAN) are identical to UPC identifiers except for the number of digits that they include.

There are two versions of EAN. Standard EAN (also known as EAN-13) has 10 numeric characters, two or three flag characters which identify the country of the EAN international organization that issued the number, and a checksum. Otherwise, it is identical to UPC Version A. JAN is identical to EAN-13, except for the fact that the flag field is set to 49, one of the codes assigned to Japan.

Following is a partial list of country codes:

00–13	United States and Canada
20–29	Reserved for local use (store/warehouse)
30–37	France
45, 49	Japan
46	Russian Federation
50	United Kingdom
54	Belgium and Luxembourg
57	Denmark
64	Finland
70	Norway
73	Sweden
76	Switzerland
80–83	Italy
84	Spain
87	Netherlands
90–91	Austria
93	Australia
94	New Zealand
99	Coupons

400–440	Germany
471	Taiwan
474	Estonia
475	Latvia
477	Lithuania
479	Sri Lanka
480	Philippines
482	Ukraine
484	Moldova
485	Armenia
486	Georgia
487	Kazakhstan
489	Hong Kong
520	Greece
528	Lebanon
529	Cyprus
531	Macedonia
535	Malta
539	Ireland
560	Portugal
569	Iceland
590	Poland
594	Romania
599	Hungary
600–601	South Africa
609	Mauritius
611	Morocco
613	Algeria
619	Tunisia
622	Egypt

625	Jordan
626	Iran
690–692	China
729	Israel
740–745	Guatemala, El Salvador, Honduras, Nicaragua, Costa Rica, and Panama
746	Dominican Republic
750	México
759	Venezuela
770	Colombia
773	Uruguay
775, 785	Peru
777	Bolivia
779	Argentina
780	Chile
784	Paraguay
786	Ecuador
789	Brazil
850	Cuba
858	Slovakia
859	Czech Republic
860	Yugoslavia
869	Turkey
880	South Korea
885	Thailand
888	Singapore
890	India
893	Vietnam
899	Indonesia
955	Malaysia
977	ISSN (*International Standard Serial Number* for periodicals)

978 ISBN (*International Standard Book Number*)
979 ISMN (*International Standard Music Number*)
980 Refund receipts

UCC/EAN-128

Code 128 is an alphanumeric, variable-length product identification scheme that can be scanned bidirectionally. Because of its high content density, it can encode the entire 128 ASCII character set as well as four nondata characters. The symbol, shown in Figure 1-12, includes a quiet zone, a start character, a significant amount of encoded data, a checksum character, a stop character, and a terminating quiet zone. Each data character comprises 11 black or white modules, while the stop character comprises 13 modules. A special start code pattern designates when the symbol complies with UCC/EAN system standards.

UCC/EAN-128 establishes a standard package-labeling scheme that can encode significantly more information than a product code, including such data as expiration dates and product batch numbers. However, it should be noted that this system is not designed to be used for product identification at the point of sales in retail outlets, but rather as a more robust identification scheme for supply chain management applications.

UCC/EAN-128 identifiers have two key fields: a data field, with its application identifier (AI), and the barcode symbol that identifies the encoding scheme for the data, in this case 128.

FIGURE 1-12 The UCC/EAN-128 product code, a high-density code with the ability to encode significantly more information than a standard product code.

THE APPLICATION IDENTIFIER

The AI is a prefix that identifies the meaning of data that follows and the format of the data. AIs exist for traceability, identification, locations, measurements of certain data, and other pertinent items. Examples are shown in Table 1-1 below.

TABLE 1-1 UCC Application Identifiers

DATA CONTENT	AI	PLUS THE FOLLOWING
Serial Shipping Container Code	00	Exactly 18 digits
Shipping Container Code	01	Exactly 14 digits
Batch Number	10	Up to 20 alphanumerics
Production Date (YYMMDD)	11	Exactly 6 digits
Packaging Date (YYMMDD)	13	Exactly 6 digits
Sell by Date (YYMMDD)	15	Exactly 6 digits
Expiration Date (YYMMDD)	17	Exactly 6 digits
Product Variant	20	Exactly 2 digits
Serial Number	21	Up to 20 alphanumerics
HIBCC Quantity, Date, Batch, and Link	22	Up to 29 alphanumerics
Lot Number	23*	Up to 19 alphanumerics
Quantity Each	30	
Net Weight (Kilograms)	310**	Exactly 6 digits
Length, Meters	311**	Exactly 6 digits
Width or Diameter (Meters)	312**	Exactly 6 digits
Depths (Meters)	313**	Exactly 6 digits
Area (Square Meters)	314**	Exactly 6 digits
Volume (Liters)	315**	Exactly 6 digits
Volume (Cubic Meters)	316**	Exactly 6 digits
Net Weight (Pounds)	320**	Exactly 6 digits
Customer Purchase Order Number	400	Up to 29 alphanumerics
Ship to (Deliver to) Location Code using EAN-13 or DUNS Number with leading zeros	410	Exactly 13 digits

TABLE 1-1 UCC Application Identifiers (*Continued*)

DATA CONTENT	AI	PLUS THE FOLLOWING
Bill to (Invoice to) Location Code using EAN-13 or DUNS Number with leading zeros	411	Exactly 13 digits
Purchase from	412	Exactly 13 digits
Ship to (Deliver to) Postal Code within single postal authority	420	Up to 9 alphanumerics
Ship to (Deliver to) Postal Code with 3-digit ISO Country Code Prefix	421	3 digits plus up to 9 alphanumerics
Roll Products — width, length, core diameter, direction, and splices	8001	Exactly 14 digits
Electronic Serial number for cellular mobile phone	8002	Up to 20 alphanumerics

AIs contain information that can be used to identify a *shipped item* (AI 01), including the EAN item identification number; a *logistic unit* (AI 00), which identifies the *serial shipping container code* (SSCC) and gives companies the ability to identify individual logistically managed items such as containers, pallets, sealed drums, and rolls or bags of product; and a *returnable item* such as a reusable container, in which case the code would indicate the EAN number of the item plus an optional serial number.

AIs also allow products to be traced through the supply chain. Information represented includes batch and lot numbers, serial numbers, production dates, expiration dates, and a maximum durability date. Quantity and measurement data, such as length, weight, height, and density, can also be encoded in the AI date. Such information not only smoothes wrinkles in the supply chain but is also extremely useful to warehouse managers who must move and store the products.

Transaction identifiers and location numbers that aid in delivery, ordering, and invoice reconciliation can also be encoded in the UCC/EAN-128 system by storing point of origin, port of entry, and other shipment data that is important to cargo handlers, trucking companies, and customs officials. AIs in this category might include purchase order numbers, invoicing information, location data, and ship-to information.

UCC/EAN-128 is one of the most complete alphanumeric encoding schemes available today. Powerfully robust, its barcodes include an initial nondata character known as *function 1* (FNC 1), which follows the start character. It allows scanners and processing software to discriminate between UCC/EAN-128 and other barcode schemes.

BEYOND GROCERIES

The use of barcodes has gone well beyond the sale of retail products. Researchers have barcoded bees (see Figure 1-13) to track their movements and mating habits, while the military has found other uses for identifiable insects: It uses them to find land mines (for reasons that are beyond me, bees are attracted to the smell of dynamite). Shipping companies like UPS and Federal Express apply specially designed barcodes to packages. In hospitals, patients, drugs, and medical paraphernalia all sport barcodes. The list goes on and on.

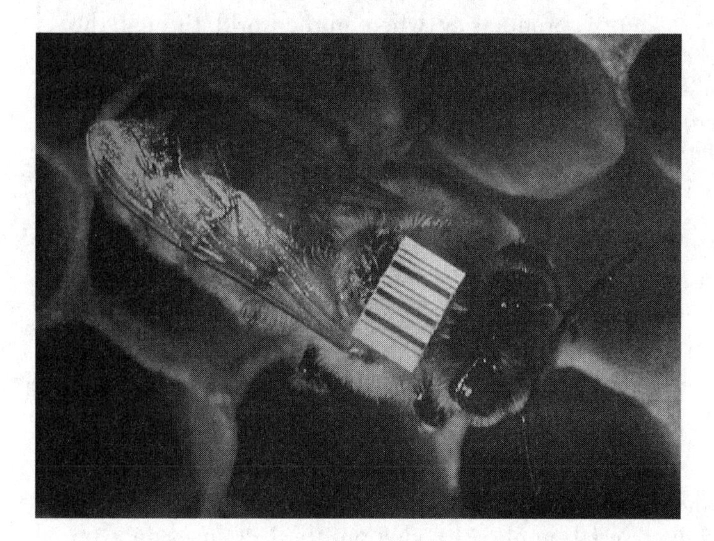

FIGURE 1-13 A barcode, which is harmlessly affixed to the top of a honeybee's thorax, allows scientists to study the complex movement of hive animals.

As barcode use has advanced, so too has the complexity and capability of the system. The EAN system, created by Woodland, has an additional pair of digits and is becoming the world's most widely deployed inventory tracking system. Other more specialized systems have evolved that can display letters in addition to numbers.

SO . . . WHAT?

The reader may wonder why we have spent so much time discussing the birth of the barcode, given that this is a book about RFID. The reason is one of evolution. Barcodes were created to solve the problem of managing large complex inventories. As markets grew and became more diverse, and as they changed from local to global, a need emerged for a mechanized system that would simplify the process of managing complex invento-

THE AUTO-ID CENTER

The Auto-ID Center, which ceased to exist in October 2003, was a nonprofit organization established by MIT to develop an Internet-based system to identify products anywhere in the world, through the use of an *electronic product code* (EPC). The organization was funded by a large number of corporations interested in the use of RFID and open standards to track products and services throughout the supply chain. In October 2003, the organization's administrative functions ended, while the research functions became Auto-ID Labs.

Replacing the Auto-ID Center is EPCglobal, an alliance between EAN International and the *Uniform Code Council* (UCC). EPCglobal is a nonprofit industry-based organization that operates with the stated goal of establishing and supporting the use of EPC technologies and networks as the global standard for immediate, automatic, and accurate product identification. The EPCglobal network will continue to work closely with Auto-ID Labs to develop EPC technology.

The Auto-ID Center originally proposed a universal standard for product identification called the EPC. Similar to a barcode, the EPC contains numeric fields that identify the manufacturer, product, version, and product serial number. In addition, the EPC uses an extra set of digits that can identify individual, unique items.

THE ELECTRONIC PRODUCT CODE (EPC)

To accelerate the adoption of the EPC, the Auto-ID Center adopted the *Global Trade Item Number* (GTIN) standards. The GTIN is a 14-digit, globally unique number that identifies products and services using existing identification codes. These include all of the major product standards including UCC-12 (UPC), EAN/UCC-13 (encoded as an EAN symbol), EAN/UCC-14 (encoded as UCC/EAN-128 or ITF symbols), and UCC/EAN-8 (also encoded as an EAN symbol). It is important to note that adoption of the GTIN does not affect the ongoing use of existing product identification standards. It is an umbrella group under which virtually all existing barcodes fall.

The EPC comprises a header followed by three data fields. The header identifies the size of the data fields that follow (the standard is designed to support 64- and 96-bit EPCs). The first field identifies the manufacturer; the second, the product; and the third, the item's unique serial number. By encoding the data as a series of fields, a scanner can conduct customized search routines, such as by manufacturer or item type.

The EPC design is rather ingenious and is based on the *Domain Name Service* (DNS) used in the Internet to locate an address. The first key element is called the *Object Name Service* (ONS). ONS directs a querying computer to an IP address where data about a particular product is archived, typically a server. Once located, the database content can be used to create reports, manage inventory, and so on. Closely related to the data itself is the technique crafted by the Auto-ID Center known as the *Physical Markup Language* (PML). PML is based largely on the *eXtensible Markup Language* (XML), which is becoming the de facto standard that describes data types. Under the terms of the EPC architecture, PML files will reside on PML servers that are universally accessible.

ries. The barcode system was the first successful model for mechanized inventory control.

As barcodes became more common in the 1980s and 1990s, challenges began to present themselves. Barcode systems were designed to operate in environments where the purchased product with its barcode label would necessarily pass in close proximity to the reader. This model proved to be adequate for retail environments such as grocery and hardware stores, certain assembly line environments, and specialized applications such as the tracking of train cars. However, as scanning technology became more commonplace and cost effective, new uses for it began to emerge. Numerous opportunities presented themselves as possible new applications. As new applications emerged, however, they brought with them a new set of challenges. For example some applications demanded a greater working distance between the barcode on the product and the reader in the work environment. Unfortunately the laser-based devices were designed to operate over short distances. Some readers, such as the handheld devices commonly seen in retail stores, could operate over distances of several feet. However, they required careful aiming on the part of the user.

As we will discuss in the next chapter, RFID came about as a way to overcome some of the perceived shortcomings of UPC barcodes. As we now know, the UPC identifies a specific type of

> Another challenge that emerged with the use of barcodes was that of information storage capacity. Standard barcodes could represent a limited amount of information, and while the information they carried was adequate for most applications, new applications began to emerge that demanded more. Grocers, for example, wanted more information available to them about the perishable food items that they purchased such as expiration dates, shipment dates, and points of origin. Others saw value in a coding scheme that would allow them to capture additional information at various points as a way of enhancing supply chain management. The UCC/EAN-128 product identification scheme extends the use of the standard barcode, although not for retail applications. Some organizations created specialized barcodes for their own use, which were capable of encoding additional information to satisfy their requirements. These systems, however, were proprietary, and therefore of little use to the general marketplace. Furthermore, any change or augmentation to the existing system would result in significant cost to users because of the likely need to modify scanning equipment and the data collection software. And while this might be a reasonable expectation in some situations, it would not be acceptable to the mass market. A better solution was required.

product. The code on a candy bar, for example, identifies the type of candy bar, but does not identify the particular candy bar that you're eating. The next development for product identification had to overcome this perceived shortcoming of existing systems. Furthermore, because of the dynamic nature of the supply chain environment, it became clear that a more flexible system that would allow the information to be dynamically changed had significant value for buyers and sellers alike.

CONCLUSION

Consider now the role that product identification—barcodes, for the most part—plays in the successful execution of supply chain management. Barcodes are printed on virtually every product that is sold today, allowing them to be quickly identified by a checkout scanner or at any point along the supply chain. Quantities of the same product are marked with a barcode, either stamped on a pallet or printed on the box they are shipped in, facilitating their ability to be tracked as they pass through intermediate warehouses and shipping vehicles. Some of these vehicles, including train cars and trailers, are marked with their own barcodes so that they themselves can be readily identified and tracked. Clearly the ability to identify products by quickly scanning a label is an advantage, but there are also disadvantages inherent in the barcode system. First, they are limited in terms of the amount of information they can carry. Second, they are far too easily smudged, damaged, or lost. Third, and perhaps most important, they require physical scanning and must therefore be visible and accessible to the scanning device, which implies that the package on which they are printed must be oriented in a particular way as it passes through the scanner, a process that often involves human intervention—and therefore added cost.

These factors, limited information-carrying capacity, physical vulnerability, and the need for a human in the loop, all lead to the conclusion that a better technology is required in modern supply chains. That technology is RFID.

RFID HISTORY

Before we dive into the inner workings of RFID technology, it is first important to understand where it came from. RFID has its roots in early military identification systems, and is based on an array of technological innovations that began in the early 1940s. The work that is most often cited as the first insight into the potential of RFID is Harry Stockman's "Communication by Means of Reflected Power," a paper published in the October 1948 issue of *Proceedings of the IRE*. In the paper, he discusses the use of a reflected radio signal as a way to identify a remote object based on the reflection signature from the object. Other papers followed in the early 1950s: D.B. Harris's *Radio Transmission Systems with Modulatable Passive Responder* and F.L. Vernon's *Application of the Microwave Homodyne*. Both spoke of systems in which a transmitted radio signal would yield an identifiable, measurable, and recognizable return signal. This technique was a well-known phenomenon at the time because of the growing use of *radio detection and ranging* (radar).

PRE-RFID: THE ARRIVAL OF RADAR

Radar was the technological precursor to RFID, and radar owed its beginnings to the work done by such early luminaries as Heinrich Hertz, the German researcher who identified and studied the wave-based nature of radio. And while Hertz, Guglielmo Marconi, and other early radio researchers played significant roles in the development of the technology, no single person deserves credit for the development of radar.

Prior to the invention of radar, a great deal of effort went into understanding the behavior of radio waves, and a great deal of that research was performed by Hertz. One result of his work was the observation that some radio waves passed through solid objects, while others were reflected by them. As a result of his work, he developed a technique for measuring the velocity of these reflected waves, and was therefore able to develop a technique for measuring the distance between himself (the origin of the reflected energy) and the object that reflected it. Early on, he presented a crude device to the German Navy, demonstrating its strategic value as a means of detecting other vessels, whether as a defensive measure or for rescue activities. They showed very little interest, and he dropped his relationship with them soon afterward.

One of the major drivers for continued work on radar technology, other than the pure science that developed from Hertz's work, was the demise of the Titanic in 1912. Scientists saw tremendous value in a system that would allow a seaborne vessel to detect unseen objects at a distance, thus preventing similar disasters. And because the transmitted signals that Hertz and his fellow researchers were using would pass through weather and darkness, they clearly had promise and were seen as something that would one day provide immense value—but not quite yet.

In the early 1920s, researchers at the United States Naval Research Laboratory were watching Hertz's efforts closely, because they, too, saw both civilian and military promise in his technology, in spite of the fact that he had not yet created a viable product based on it. The Naval Research Lab had watched Hertz's studies closely and had noticed that an interference pattern was returned when a naval ship or aircraft passed through the transmitted radio signal, as shown in Figure 1-14. Unfortunately they did not yet see a viable application, and their promise to fund further research was abandoned.

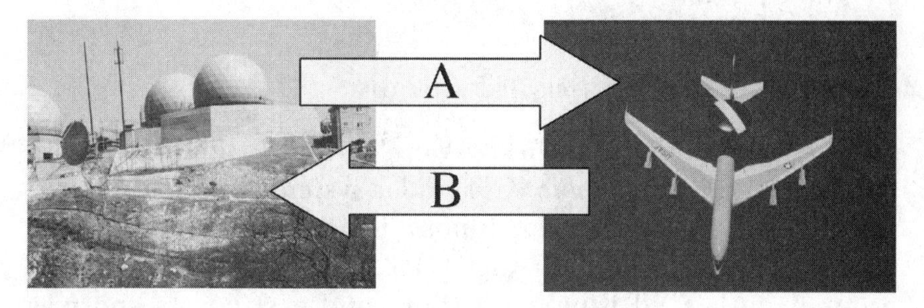

FIGURE 1-14 Radar installation transmits signal (A); aircraft reflects part of signal back (B), creating a detectable "echo."

FIGURE 1-15 Early radar monitoring installation.

It was World War II that finally gave radar its impetus for creation. As hostilities among the countries that would ultimately take active roles in the war grew, paranoia over the ability to detect an invading army, particularly at night or in foggy conditions, grew as well. Within a short period of time active radar research was underway in the United States, the United Kingdom, Germany, France, the USSR, Italy, and Japan. By the time the war began in 1939, most of them had developed deployed functional radar systems, as shown in Figure 1-15. Their independent yet simultaneous efforts gave rise to radar as we know it today.

FIRST DEPLOYMENTS: CHAIN HOME RADAR

At the beginning of World War II, the United Kingdom deployed the *Chain Home* (CH) radar system as an early warning mechanism for detecting inbound enemy aircraft and ships. The system, shown in Figure 1-16, was deployed in an arc around the United Kingdom's south and east coasts and was immediately effective except for one type of inbound enemy: low flying aircraft. This was an enormous problem since

FIGURE 1-16 Photograph of an early CH radar installation in the United Kingdom.

German planes began to bomb London and outlying areas with impunity. As a result of yeoman efforts on the part of scientists and researchers in the field, a new type of radar known as *Chain Home Low* (CHL) radar was developed and deployed in 1939. The devices were deployed along the same arc as the CH systems, but were staggered between the CH radar installations. The newly installed CHL radar systems could detect aircraft flying as low as 500 feet above the deck and at distances up to 30 miles away.

An equally urgent simultaneous effort was underway in the United States at this time. A small division of the U.S. Navy was working around the clock to develop a more sensitive radar tracking system. Working collaboratively with their U.K.-based allies, they developed a highly sensitive electron tube that could generate the necessary high-frequency, high-power radio signal required to detect targets at significantly greater distances than earlier systems such as the United Kingdom's CH and CHL deployments. Because of the high frequency (and therefore extraordinarily short wavelength) that operationally characterized the system, it was referred to as microwave radar. It was extremely accurate and could operate at equally extreme distances. The result of these advancements was significant. Defense forces with radar detection capability could now track the locations of aircraft and ships while en route, calculate their exact speed, and ensure their safety while in transit. This effort led to the development of the generic "Identify: Friend or Foe" model that in turn became the "squawk" transponder model used in modern aircraft communications today. Squawk codes are issued to outbound aircraft as part of the departure clearance procedure, as well as to aircraft entering air-traffic-controlled airspace. Also known as *secondary surveillance radar* (SSR), squawk codes are four letter codes that allow each aircraft to be uniquely identified. They are input manually into an onboard transponder, causing the aircraft to "squawk" the code when interrogated.

As often happens with technology advances that occur during wartime, civilian applications for radar were on hold until the war ended. Once it did, however, efforts to develop applications for the civilian use of radar technology moved to the forefront and the technology evolved rapidly. Today it is so accurate and capable that radar receivers can identify and extract only the desired signal, filtering out all of the spurious energy that inevitably clutters the spectrum.

And what does radar have to do with RFID? The link is interesting. Consider what RFID means: *radio frequency identi-*

Radar is based on the concept of an echo. The base station transmits a narrow, intense beam of electromagnetic energy from an antenna that is usually spinning (see Figure 1-17). The sending device then turns off its transmitter and turns on its receiver, listening for a return echo. If the transmitted beam hits an object, a fraction of the transmitted energy is returned as an echo. Based on the return signature, the radar can accurately determine the location of the object, as well as its velocity and heading.

FIGURE 1-17 An early radar tower—note the spinning send/receive device on top.

fication, the ability to identify someone or something through the receipt of some kind of radio signal. Passive RFID tags work very much like the targets that radar systems strive to detect. When a passive RFID tag is placed within the operating range of a reader, energy from the reader activates the tag, causing it to emit a signal that can be read by the reader. And while this emitted signal is not a reflected signal like those that radar systems rely on, the concept is identical: An emitted energy beam causes a return signal to be generated that can be read electronically by a scanning device.

Post-Radar Development Efforts

A flurry of research activity characterized the 1960s, as spellbinding discoveries captivated the scientific world: Modern computers came into being, the integrated circuit arrived, lasers were born, and digital data networks were perfected. A collection of fascinating studies emerged, including Robert Richardson's *Remotely Activated Radio Frequency Powered Devices* (1963) and J.P. Vinding's *Interrogator-Responder Identification System* (1967). Commercial efforts during this period were alive and prospering as well. Knogo (today a division of Sentry Technology Corporation), Sensormatic (a division of Tyco International), and Checkpoint were founded during this period, all with the intent of creating electronic article surveillance products to fight shoplifters. They initially created a 1-bit tag that made it possible to simply detect the presence of a tag as a would-be thief walked through a protected portal. If the shoplifter carried a piece of merchandise from the store without first paying for it—during which time the embedded tag would be deactivated—an alarm would sound, alerting store security.

During the 1970s, the *Los Alamos Scientific Laboratory* (LASL) in northern New Mexico became the center point for research that led to the development of modern RFID systems. In 1975 Robert Freyman, Steven Depp, and Alfred Koelle published a seminal paper with the enviable title, *Short-Range Radio-Telemetry for Electronic Identification Using Modulated Backscatter.* Based on the work done at LASL and at a scattering of universities around the world, commercial deployments of radio identification technology began to appear. In 1973, Raytheon Corporation released its "Raytag" product; RCA and Fairchild Semiconductor followed shortly thereafter with their own offerings. At roughly the same time, Schlage Electronics (now a division of Honeywell) developed and sold an analog RFID card, primarily targeted at military access control.

As products began to appear in the marketplace, customers emerged to use them. The first commercial application with merit for electronic identification technology was vehicle tracking, and the first deployment was with the Port Authority of New York and New Jersey, which tested electronic automated toll collection applications offered by Philips, GE, Westinghouse, and Glenayre Technologies. Other vendors rolled out radio detection applications for livestock tracking, vehicle movement monitoring, and supply chain management (such as it was in the 1970s).

Los Alamos continued to push the limits of the technology, working closely with not only researchers but with implementers as well. They collaborated extensively with vehicle tracking application developers and spun off a number of smaller companies to pursue commercial development of early RFID technology. One of these companies was Santa Cruz-based *Identronix* (IDX, acquired to form RFID solutions provider Escort Memory Systems); another was Amtech, later acquired by Intermec and subsequently sold to TransCore, today a leader in vehicle and package tracking technologies and applications.

During the 1980s and early 1990s, RFID systems matured, applications evolved, and the market for them stabilized. The ability to track vehicles and containers proved to be the most significant driver of continued activity, and the Association of American Railroads and the Container Handling Cooperative Program were both extremely involved in the development of both standards and applications of the technology. Both of the industries represented by these trade support groups operated like a library: Temporarily idle train cars left on sidings could be taken and used by any railroad, regardless of the name printed on the side of the boxcar. Similarly, pallets and large shipping containers were shared among industries, leading to the need to track them as they move freely around the country and the world.

THE FIRST RFID TAG: MARIO CARDULLO

In 1969 Mario Cardullo was the corporate planning officer reporting to the chairman of satellite provider Comsat. In the spring of 1969, he found himself on a plane, seated beside an IBM engineer who was at the time involved in the implementation the optical CARTRAK system that was being developed for the railroad industry. This system was based on a reflective multicolored barcode on the side of each railroad car. As a string of cars passed through a train yard, an optical reader in the station would read the barcode on each car and note its identity in a database. The problem with this system was the same problem that continues to plague barcode systems today; dirt, intense sunlight, scratches in the barcode, and other impairments often made it difficult for readers to accurately scan the identifier on each piece of rolling stock.

Following his conversation with the IBM engineer, Cardullo sketched out the idea for the first RFID-like tag with dynamic memory. His original sketch included a transmitter, a receiver, a small amount of internal memory, and a dedicated power source.

Several years later, Cardullo left Comsat and started ComServe, a firm initially dedicated to the design and manufacturing of an automated toll collection system. He had continued to refine the earlier concept, knowing that someday it would prove to be a valuable innovation with commercial appeal. On January 23, 1973, the U.S. Patent Office approved Cardullo's request for the following patent:

1. A transponder comprising:
   a. Memory [as a] means for storing data
   b. Means responsive to transmitted code signal for selective writing data into or reading data out from the memory and for transmitting as an answerback signal data read-out from the memory
2. Means for internally generating operating power for the transponder from the transmitted code signal
3. A transponder where the transmitted code signal comprises a modulated carrier wave, this being the means for generating operating power that comprises a means of detecting the carrier wave and producing an operating power output signal, responsive to the operating power output signal for powering the transponder and the modulations of the carrier wave containing data and command information
4. A transponder as defined in the patent where the carrier wave is of radio frequency

5. A transponder as defined in the patent where the carrier wave is of light frequency
6. A transponder as defined in the patent where the carrier wave is of acoustic frequency
7. An interrogation system

The patent document also detailed potential applications for its use, including an automated highway toll system similar to the E-Z Pass in widespread use today (discussed shortly), and "a transponder which would be physically small in size such that the device is highly portable, can be easily hidden, if desired, and can be carried and placed in or upon many different objects."

Based on his patent, Cardullo and his team of engineers set to work designing and building the first actual RFID tag. This proved to be a significant challenge: In the early 1970s the only nonvolatile memory was based on the large arrays of ferrite cores (Figure 1-18) used in mainframe computers. Cardullo and his engineers had already determined that their memory

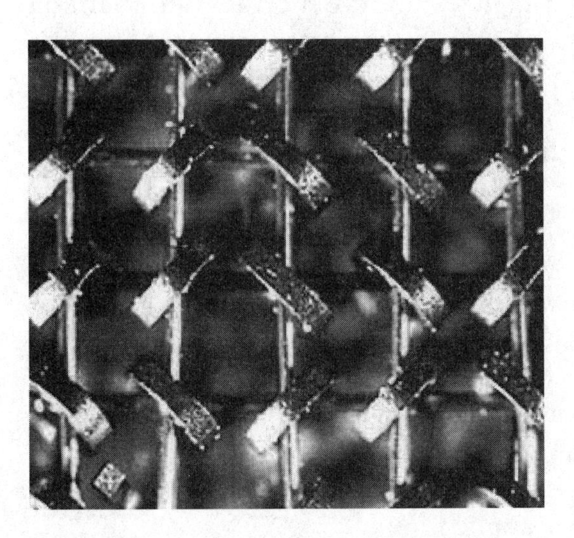

FIGURE 1-18 The ferrite cores shown here were used in early computer memory systems. The ferrite rings could be magnetized or de-magnetized to indicate binary (0/1) information.

requirements were relatively small (16 bits); they purchased the individual ferrite rings and hand wound them to create the memory arrays they needed for their design.

In 1971 ComServe met with the New York Port Authority. They demonstrated their toll collection application, at the end of which they were told "no one will ever mount those transponders on their car windows." Prophetically, based on current resistance to RFID, the port representatives expressed concerns over invasion of privacy.

Two years later, the New York Port Authority was testing the GE, Glenayre, Philips, and Westinghouse systems described earlier. Cardullo left ComServe, and shortly thereafter the firm went out of business. Nevertheless, the firm was pivotally important in securing the first generic patent for an RFID device, and when the patent expired in the 1990s, its influence was substantial.

LATER DEVELOPMENTS

As RFID continued to evolve, the energy surrounding it grew. A collection of highway agencies in the northeastern United States formed the cooperative E-Z Pass Interagency Group, the result of which is a regionwide toll collection system. Subscribers mount a small device behind their rearview mirrors, and as they pass through the toll gate their presence is detected and they are billed for road use on a monthly basis. The model was popular and soon appeared throughout the world.

Texas Instruments has become the best-known advocate of RFID technology and is one of the most active in terms of research, development, and manufacturing. Their *TI RF Identification Systems* (TIRIS) products began with vehicle antitheft devices, but now have grown to include fuel dispensing systems (the ExxonMobil Speedpass, for example, shown in Figure 1-19), keyless entry devices, ski pass control, secure entry systems, and other applications.

As RFID has become more and more accepted, a variety of companies has emerged to take advantage of the technology

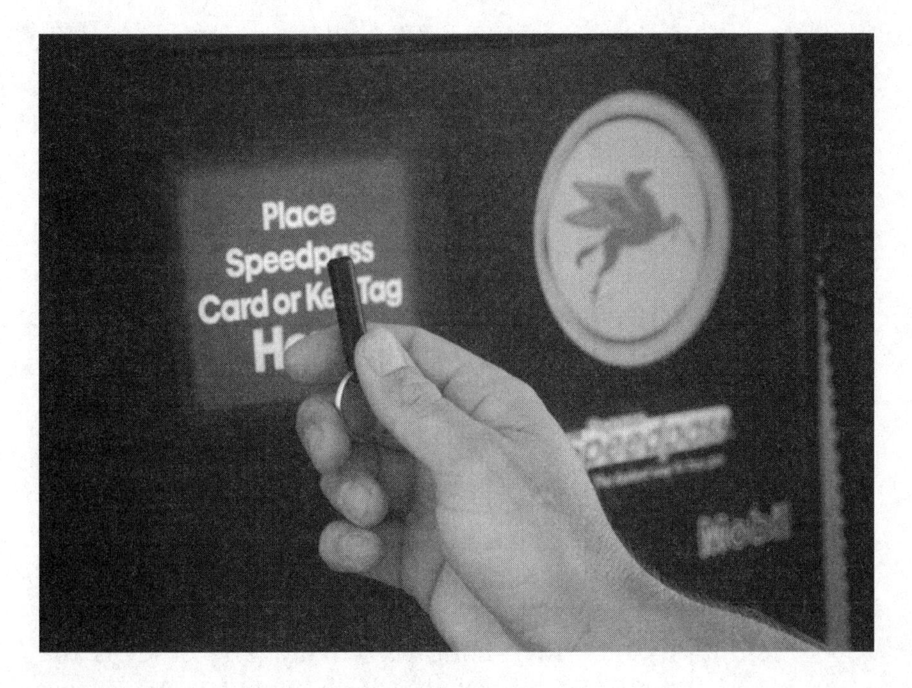

FIGURE 1-19 The ExxonMobil Speedpass.

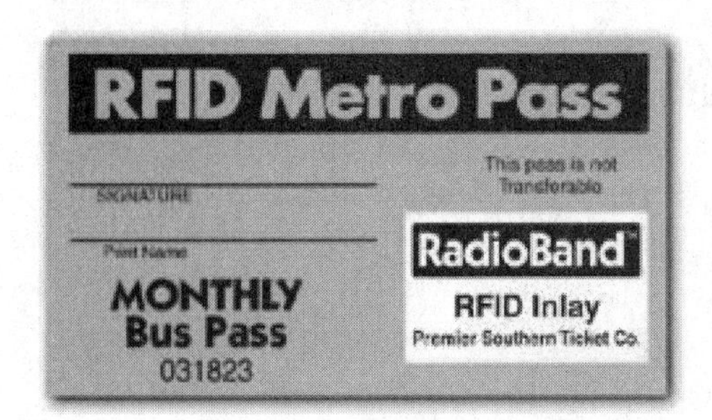

FIGURE 1-20 An RFID Metro Pass. *(Photo courtesy Premier Southern Ticket Corporation.)*

with a range of innovative products. Premier Southern Ticket, for example, manufactures RFID-equipped tickets for large group events such as carnivals, county fairs, concerts, bus and

FIGURE 1-21 An RFID Carnival Pass.

train tickets, and sporting events. Examples are shown in Figures 1-20 and 1-21.

IN SUMMARY

The evolution of RFID is based on two parallel developmental paths. One is the evolution of the underlying radio-based technology that began with Heinrich Hertz and his electromagnetic experimentation and which continues today. The other is the evolution of the application set that RFID enables. As we will see in the next chapter, RFID is a reasonably complex set of radio technologies. On the other hand, that technological complexity enables the delivery of a broad spectrum of functions that in turn enables the effectiveness of the overall supply chain, regardless of the industry in which it is implemented.

PART TWO

RFID IN DETAIL

In this section, we will discuss the inner workings of RFID systems. We begin with a discussion of the general details of RFID system anatomy, followed by detailed discussions of the components that make up a typical RFID system and the underlying technologies that make them work. We conclude with a discussion about RFID security, which has become both very visible and highly overblown in terms of its threat to society at large. On the other hand, RFID represents a new opportunity to strengthen those areas of national security that need attention; we will discuss these applications in detail.

TYPICAL RFID SYSTEM COMPONENTS

For the most part, RFID systems comprise three principal components, as shown in Figure 2-1. The first is the *transponder*, which is affixed to the item that is to be tracked or identified within the supply chain by the RFID system; the *reader*, which has a number of varied responsibilities including powering the transponder, identifying it, reading data from it, writing to it, and communicating with a data collection application; and the *data collection application*, which receives data from the reader, enters the data into a database, and provides access to the data in a number of forms that are useful to the sponsoring organization. For example, an RFID application deployed in a warehouse environment may be designed to track the movement of

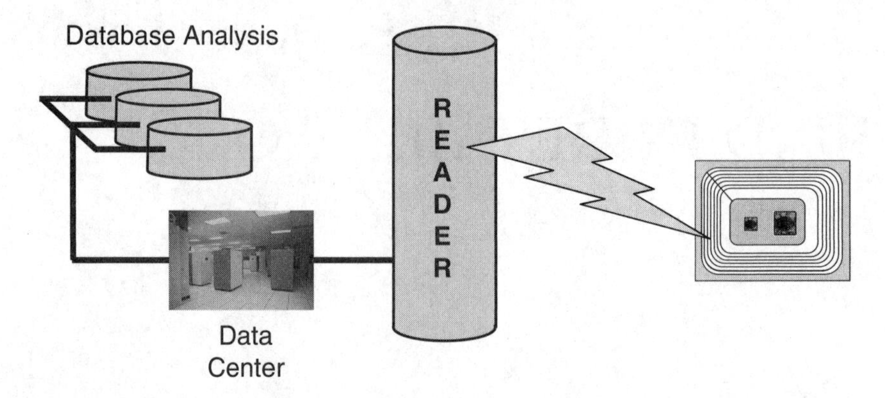

FIGURE 2-1 The reader sits between the tag, which generates the data, and the applications that analyze the data to produce management information.

pallets of product as they make their way through the warehouse. RFID transponders mounted on each pallet pass within reading distance of readers that are deployed at various places along the supply chain, and as they do so, they are energized by each reader, causing them to transmit their identification information to the reader, which in turn passes the information along to the supply chain management application. The application creates reports that detail product location, warehouse product management velocity, on-hand inventory data, efficiency reports, and a plethora of additional information that is useful to warehouse management. Because the transponders are coded to a particular pallet, and because the supply chain management system knows what products are on each pallet, the products can be tracked at the pallet level throughout the warehouse environment. The pallet-mounted RFID transponder can be recoded with different information each time the pallet is reused, thus ensuring that the technology is cost effective.

There is a fourth critical component in RFID systems: the air interface between the transponder and the reader. And although it's clearly not a component in the true sense of the word, it does require design considerations. As a result, we will discuss it in its own right in the section that follows.

TRANSPONDERS

Transponders come in a variety of forms and are often application specific. We begin our discussion with device behavior.

PASSIVE VS. ACTIVE TRANSPONDERS

RFID transponders are either *passive* or *active* devices. Passive RFID transponders do not have a dedicated power supply. Instead, they derive their operating power from the electrical field generated by the reader, and therefore do not operate unless they are in very close proximity (a few centimeters) to the reader. This limited operating distance has the added value of preventing a transponder from being inadvertently turned on, simply because its owner unwittingly walked too close to a fuel pump or other activating device. It also should assuage some of the oft-stated concerns about privacy and security. Because they do not have internal batteries, passive transponders are quite small and can be installed unobtrusively. Figure 2-2 is an example of a passive transponder that can easily be affixed to the

FIGURE 2-2 A passive transponder is affixed to an adhesive plastic sheet, which can be easily inserted in a book, tape, or other item subject to theft.

inside cover of a book or other small retail product. It is small, lightweight, and virtually invisible.

Active transponders, on the other hand, such as that shown in Figure 2-3, *do* have an internal battery and therefore have significantly greater read range, as much as 50 feet in some cases. They are, however, significantly larger than their passive counterparts and therefore lend themselves to a different set of applications. For example, they are often used for automated toll-paying applications (EZ Pass, for example) or for tracking large items in a warehouse (Figure 2-4), such as a product-laden pallet. The choice of a passive vs. an active transponder is purely dependent upon the application for which it will be used. Passive transponders are typically used in situations where large numbers of them will be required for a successful deployment (they're significantly less expensive than active devices), where the application is such that the transponder can be placed in close proximity to the reader as in a fuel-dispensing application (Figure 2-5), a bookstore/library access-control application (Figure 2-6), or the like. Active devices, on the other hand, are

FIGURE 2-3 An active transponder. Because it is battery-powered, the read range of this device is significantly greater than a passive device. (*Photo courtesy Texas Instruments.*)

FIGURE 2-4 An active RFID tag on the side of the crate allows it to be tracked throughout the supply chain at a significant distance from the reader.

FIGURE 2-5 Dispensing fuel using an RFID-based Speedpass.

FIGURE 2-6 Access control using RFID in a library.

FIGURE 2-7 Smaller devices that can be attached to small product-level items.

often chosen for situations where they will be deployed in smaller numbers (at the pallet or large container level rather than at the discrete product level, Figure 2-7) and in situations where the transponder must necessarily be at a greater distance

FIGURE 2-8 Active RFID can be used to identify individual containers in a shipping environment.

from the reader, such as in a highway toll-taking application, a trucking or shipping container identification application (Figure 2-8), or certain supply chain implementations.

OPERATING FREQUENCIES

RFID transponders and readers operate within several distinct frequency ranges, each of which is intended for specific application characteristics. *Low-frequency devices* (30 to 300 KHz) are typically found in passive tags and used in short-range applications such as livestock identification (Figure 2-9) and for antitheft systems in automobiles (Figure 2-10). A typical device would operate between 125 and 134 KHz.

High-frequency devices (3 to 30 MHz) are used most commonly in smart card and smart label applications such as baggage tracking (Figure 2-11) or small product labeling (Figure 2-12). Typical systems operate in the range of 13.56 MHz.

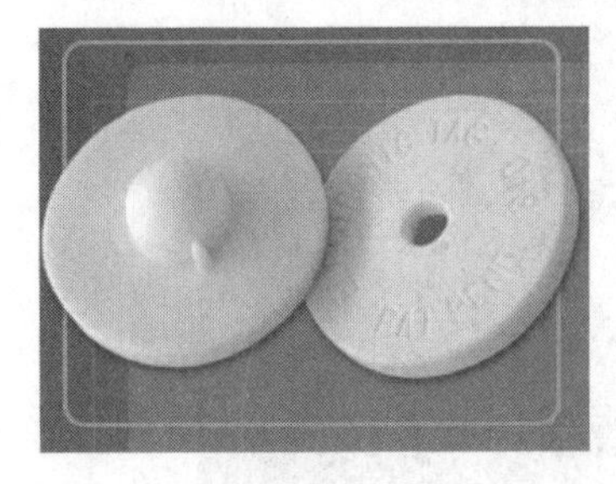

FIGURE 2-9 An ear tag used for livestock tracking. *(Photo courtesy SOKYMAT Corporation.)*

FIGURE 2-10 RFID used to implement an automobile antitheft system. The RFID tag embedded in the base of the key prevents the car from being started unless the key is in the lock—and the transponder adjacent to the reader. *(Photo courtesy Texas Instruments.)*

FIGURE 2-11 An RFID baggage label, now being deployed in major airports for secure and accurate baggage tracking. *(Photo courtesy Texas Instruments.)*

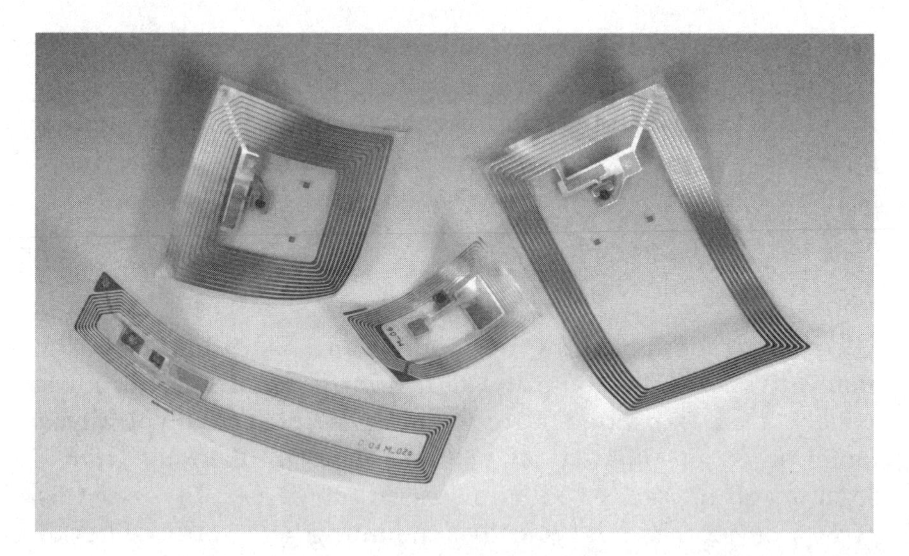

FIGURE 2-12 Sticky RFID labels can be affixed to a variety of individual small products. *(Photo courtesy Texas Instruments.)*

Very high-frequency devices (300 MHz to 3 GHz) are primarily used in highway toll-collection applications. In the United States, systems typically operate at 900 MHz or 2.45 GHz; in Europe, similar systems operate in the 5.8 GHz range.

FREQUENCY ATTRIBUTES

There are definite tradeoffs among the various RFID operating frequencies. Lower-frequency systems (typically below 500 kHz) require minimal operating power, are relatively inexpensive, have transmission capabilities in the short-to-medium range, and support reasonable date rates. They are not sensitive to orientation, can usually be read through metallic overlays, but are somewhat noise sensitive.

Higher-frequency systems, operating above 1 MHz, require higher power and are somewhat more expensive than their lower-frequency counterparts. They support greater read distances as well as higher data transmission rates, and are orientation sensitive. For example, one fuel-dispensing option, mounted in the back window of an automobile, must be oriented at a right angle to the pump to be properly detected. Higher-frequency systems will not penetrate metallic surfaces but are also less prone to noise interference than are lower-frequency systems.

FORM FACTORS

RFID transponders come in a variety of shapes and sizes for many diverse applications, as shown in Figure 2-13. The most widely used transponder form factor is the *disk,* typically a small, injection-molded device that ranges in diameter from a fraction of an inch to four inches (see Figure 2-14). For this type of device, there is typically a mounting hole in the center of the disk so that it can be affixed to a pallet or other transport container. Other disk-shaped transponders are extraordinarily small, no larger than a shirt button, and are designed to be sewn into apparel or affixed to a cloth label for product identification purposes.

A second common form factor is the molded styrene or epoxy resin device. These devices are designed to be small and

FIGURE 2-13 RFID tags come in a variety of shapes and sizes, as shown above.

FIGURE 2-14 An RFID disk-shaped transponder that can be affixed to a pallet or other large shipping container.

FIGURE 2-15 The Speedpass form factor, designed to hang on a keychain and used to trigger a fuel pump.

sturdy and are often packaged as keychain fobs such as the Speedpass shown in Figure 2-15. RFID transponders are also molded into the plastic heads of some automobile keys; the reader, mounted in the steering column, will not allow the car to start unless the proper RFID signal is detected, thus serving as an effective theft deterrent. Similarly, the small rectangular form factor (sometimes referred to as a wedge) is commonly seen in factory floor environments, supporting supply chain management applications. An example is shown in Figure 2-16. Because these devices are wedge-shaped, they can easily be oriented for those applications that require precise tag orientation.

One of the most intriguing forms of RFID transponder is the glass tube, shown in Figure 2-17. These devices are very

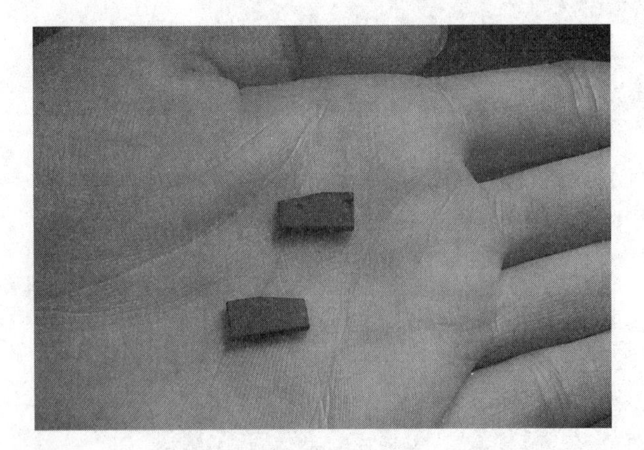

FIGURE 2-16 A wedge-shaped transponder: The irregular shape allows it to be oriented a particular way to ensure maximum signal strength exists between the reader and the transponder.

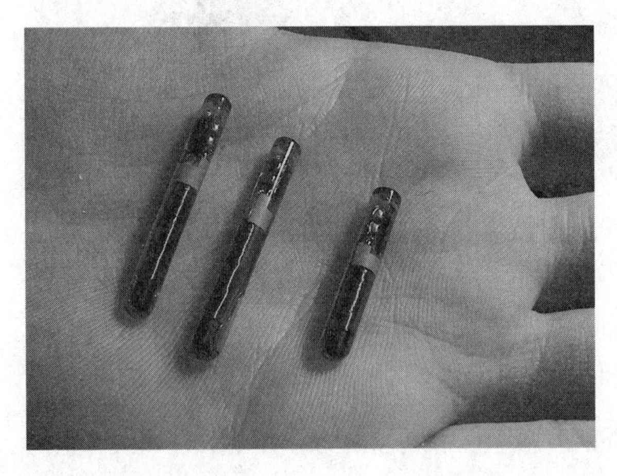

FIGURE 2-17 Glass transponders, designed for subcutaneous injection.

small, roughly one-eighth inch in diameter and less than an inch long, and are designed for subcutaneous injection in livestock. As long as the device is sterile at the time of injection, it can be mechanically introduced under the skin using a large-bore hypodermic designed for the task, without undue risk or pain to the animal. These devices have become more important and visible in the last year as outbreaks of bovine spongiform encephalopathy (mad cow disease) have occurred. When the Washington state outbreak occurred at the end of 2003, it took public health officials several weeks to determine the genetic origin of the animal. Had the information been encoded on an internal RFID transponder, injected at the time of the calf's birth, the animal's origin could have been determined in a matter of seconds. Clearly, mad cow disease represents a compelling argument for the universal implementation of RFID tagging of all livestock.

There are other forms of animal transponders including collar tags, ear tags (Figure 2-18), and ceramic tags that are designed to be swallowed by the cow and to remain permanently within the animal's rumen. Injectable tags are manufactured by literally dozens of companies, including Biomark,

FIGURE 2-18 RFID ear tags for livestock identification. *(Photo courtesy RFID Journal.)*

which specializes in fish tags such as the one being injected into the large snake shown in Figure 2-19. Other examples include ID cards (Figure 2-20), sticky labels (Figure 2-21), and even tiny rings (Figure 2-22) designed to be attached to a pigeon's leg. When a bird takes part in a race, the RFID transponder

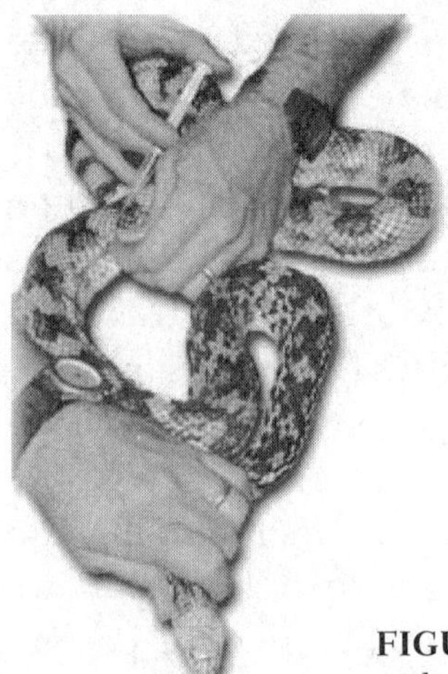

FIGURE 2-19 Snake being subcutaneously tagged with injectable RFID tag. *(Photo courtesy Biomark, Inc.)*

FIGURE 2-20 Among others, RFID identification cards (right foreground) are becoming commonly used for secure facility access. *(Photo courtesy Texas Instruments.)*

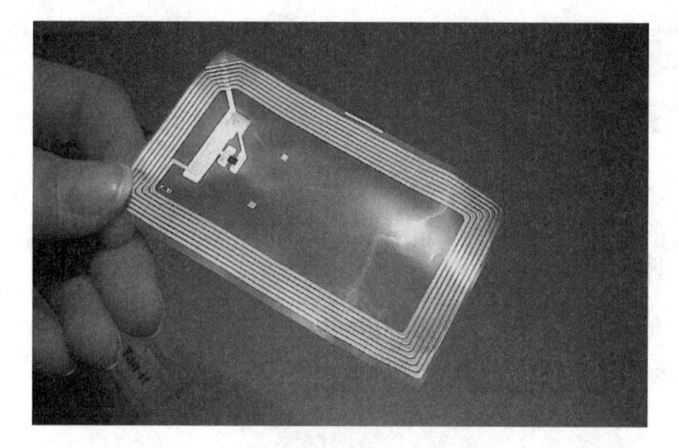

FIGURE 2-21 A low-cost sticky RFID label.

FIGURE 2-22 RFID is even used to track racing pigeons. These rings are attached to the birds' legs, providing precise arrival information to race coordinators. *(Photo courtesy SOKYMAT Corporation.)*

secreted inside the leg band helps to recognize the exact time the bird arrives at the finish line. Amazing. Finally, there are cylindrical transponders such as the one shown in Figure 2-23, which are designed to be pole-mounted and are packaged in materials that resist heat and corrosive compounds, including the stomach acids of cows.

FIGURE 2-23 Cylindrical RFID transponder, designed to be pole-mounted: The plastic housing can be used in hot or chemically hostile environments.

SMART CARDS

In 1988, the *International Organization for Standardization* (ISO) and the *International Electrotechnical Commission* (IEC) created a *Joint Technical Committee on Information Technology* (ISO/IEC JTC1), known commonly as JTC1. JTC1 consists of approximately 20 subcommittees on all areas of information technology. *Subcommittee 17* (SC17) is responsible to develop standards for identification cards and for personal identification. These standards include test methodologies, physical device characteristics, embossing, characteristics of magnetic stripe integrated circuit cards with contacts, contactless integrated circuit cards, optical memory cards, machine-readable travel documents such as passports and visas, driver's licenses, ID cards, and a variety of other personal identification modalities. They are also standardized through various ISO committees, including ISO 10536 (close-coupling smart cards), ISO 14443 (proximity-coupling smart cards), and ISO 15693, (vicinity-coupling smart cards). Each will be discussed in turn.

SC17 operates under the directives issued by JTC1 and has a number of standing *working groups* (WGs) that address the issues associated with the identification technologies. These committees are as follows:

WG1—Physical Characteristics and Test Methods for Identification Cards

WG3—Machine Readable Travel Documents

WG4—Integrated Circuit Cards with Contacts

WG5—Registration Management Group

WG7—Financial Transaction Cards

WG8—Contactless Integrated Circuit Cards, Related Devices, and Interfaces

WG9—Optical Memory Cards and Devices

OWG *(Operating Working Group)*—Technology Co-Existence on Identification Cards

WG10—Motor Vehicle Driver Licenses

WG8 is responsible for contactless smart cards, sometimes called ID-1 form factors.

Smart cards, sometimes called contactless smart cards, look (for the most part) like credit cards. Examples are shown in Figure 2-24. Because they are fairly large, there is ample room for a large internal coil (Figure 2-25), which increases the read

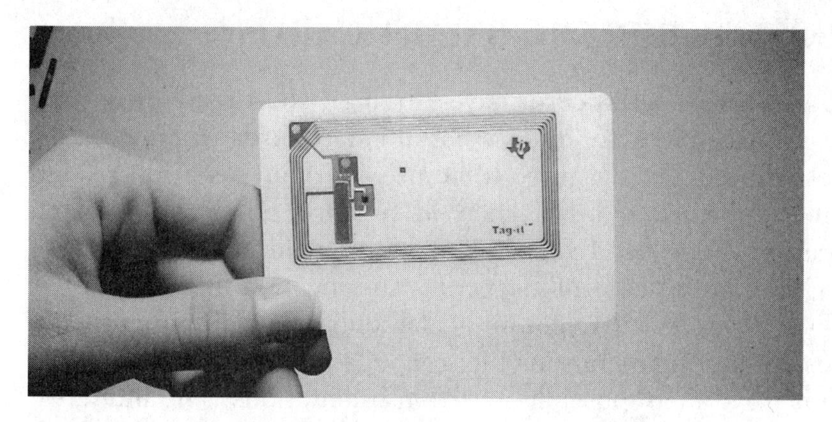

FIGURE 2-24 An RFID-based secure ID card.

FIGURE 2-25 A secure ID card, showing large internal coil.

range of card-based designs. Companies that issue contactless smart cards often print logos or designs on the card stock, and because they are of a size that is for the most part universally known, most users simply carry them in their wallets. As was mentioned earlier, three forms are defined: close-coupling smart cards, proximity-coupling smart cards, and vicinity-coupling smart cards.

CLOSE-COUPLING SMART CARDS

Close-coupling smart cards have extremely short-read ranges and are in fact similar to contact-based smart cards (Figure 2-26). These devices are designed to be used with an insertion reader, similar to what is often seen in modern hotel room doors. Because of the need for short-distance read actions, the ISO 10536 standard carefully specifies the location and size of the inductive or capacitive coupling elements. The component locations are carefully chosen (Figure 2-27) to ensure that the card can be read in all possible insertion orientations. The inductive coupling elements shown in the diagram generate the power

FIGURE 2-26 Contact-based smart card; note the contacts (arrow) on the card's surface.

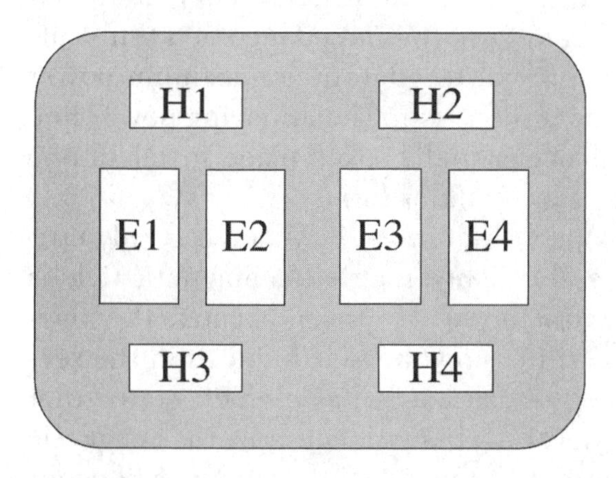

FIGURE 2-27 Close-coupling smart card showing positions of inductive (H1-H4) and capacitive (E1-E4) components.

required to operate the device, generating an alternating field of 4.9152 MHz. Coils H1 and H2 are wound in opposite directions, as are H3 and H4, so that if power is applied to them simultaneously, a 180-degree phase difference will result between the magnetic fields in the reader. At no time should these cards draw more than 200 mW of aggregate power; similarly, readers should be designed to deliver 150 mW to the card from any of the generating magnetic fields.

Data transmission from the card to the reader (we assume a read-only card in this case) can be either inductive or capacitive, but once a transmission has begun using one technique, the other cannot be used during transmission.

A BRIEF ASIDE: INDUCTIVE VS. CAPACITIVE COUPLING

Let's take a moment to define these terms because we're throwing them around like beads at Mardi Gras. *Inductive coupling* refers to the transfer of electromagnetic energy from one circuit to another as a result of the mutual *inductance* between the circuits. *Inductive coupling* may be intentional, such as in an impedance matcher that matches the impedance of a transmitter or a receiver to an antenna to guarantee maximum power transfer, or it may be unplanned, as in the annoying power line inductive coupling that occasionally takes place in telephone lines, often referred to as *crosstalk* or *hum*.

Consider the diagrams in Figure 2-28. When a steady current flows through the coil (A) shown at left, a magnetic field is induced in the coil (B) to its right. However, because the magnetic field is in a steady state, no voltage is induced in the second coil, a consequence of Faraday's Law, which states that voltage is created whenever a change occurs in the magnetic field. However, if the circuit is opened (C), a change *will* occur in the magnetic field in the right-hand coil and voltage will be induced (D).

A coil is known as a reactionary device, and it doesn't like change. When voltage is induced in coil (A), the current flows in coil (B) as it attempts to maintain the magnetic field that was already present. The induced field *always* opposes the change.

When the switch is closed again (E), allowing current to flow, an induced current in the opposite direction will once again oppose the buildup of a new magnetic field (F). This ongoing creation of voltage, which opposes the changing mag-

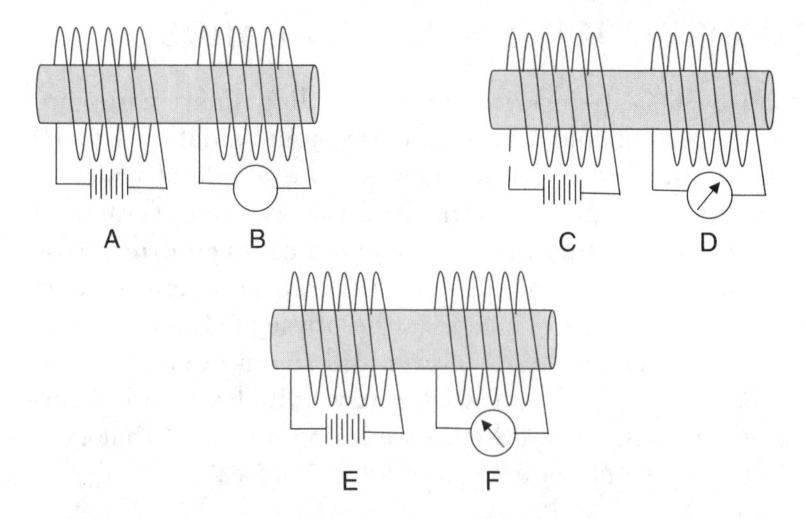

FIGURE 2-28 Induced current flow in a coil.

netic field is the basis for the device used in power supplies and RFID tags and is called a *transformer,* because the induced current flow in one coil (sometimes called the *primary*) causes a transformation in the state of one or more adjacent coils, called *secondaries*. The tendency of a change in the current of one coil to affect the current and voltage in a second coil is called *mutual inductance*. When voltage is produced because of a change in current in a coupled coil, the effect is mutual inductance. The voltage always opposes the change in the magnetic field produced by the coupled coil.

Capacitive coupling, on the other hand, is the transfer of electromagnetic energy from one circuit to another through mutual capacitance, which is nothing more than the ability of a surface to store an electric charge. *Capacitance* is simply a measure of the electrical storage capacity between the circuits. Similar to the inductive coupling phenomenon described earlier, capacitive coupling can be both intentional and unplanned.

One key difference between the two: Capacitive coupling favors the transfer of higher-frequency signal components, and inductive coupling favors lower-frequency elements.

PROXIMITY-COUPLING SMART CARDS

Governed by ISO 14443, proximity-coupling smart cards are designed to be readable at a distance of approximately 4 to 10 inches from the reader. These devices, an example of which is shown in Figure 2-29, are often used for sporting events and other large public gatherings that require access control across a large population of attendees. Like all of the contactless smart card standards, ISO 14443 defines the physical characteristics of the device (Part 1), the RF power and signal interface parameters (Part 2), the protocols that govern initialization and collision control (Part 3), and the transmission protocols that exist between the transponder and the reader (Part 4).

PART 1: PHYSICAL DEVICE PARAMETERS

Part 1 defines the physical device parameters, including size, as described in ISO standard 7810: a card that measures 3.37 inches by 2.12 inches by 0.03 inches. Because of the typical application for the physical form factor, bracelets, armbands, etc., Part 1 also details requirements for dynamic bending and torsion stress loads as well as functional limits for ultraviolet, electromagnetic, and x-ray irradiation that could occur in controlled environments where these devices could be deployed.

FIGURE 2-29 A proximity RFID card, used (in this case) for entry to an amusement park.

PART 2: RF CHARACTERISTICS

Proximity-coupling smart cards rely on a fairly large coil (see Figure 2-30) to ensure that the generated signal can be read properly at its maximum operating distance of approximately 10 inches. Power for inductively coupled smart card transponders comes from the 13.56 MHz alternating magnetic field generated by the reader (Figure 2-31).

In the technology industry, we often observe that the nice thing about standards is that there are so many to chose from. In the world of RFID transponders this is certainly true. During the early stages of RFID standards development, there was ample disagreement about the required nature of the communications interface between the transponder and the reader, the result of which was the creation of two separate standards, ISO 14443 Type A and ISO 14443 Type B. In most cases, proximity-coupled smart cards support one or the other; readers, on the other hand, must support both. To accommodate the existence of both transmission standards, readers must be able to alternate between the two recognized polling standards during the WAIT FOR COMMUNICATION FROM THE TRANSPONDER idle period. Readers cannot, however, switch between the standards during an ongoing transmission session with a device.

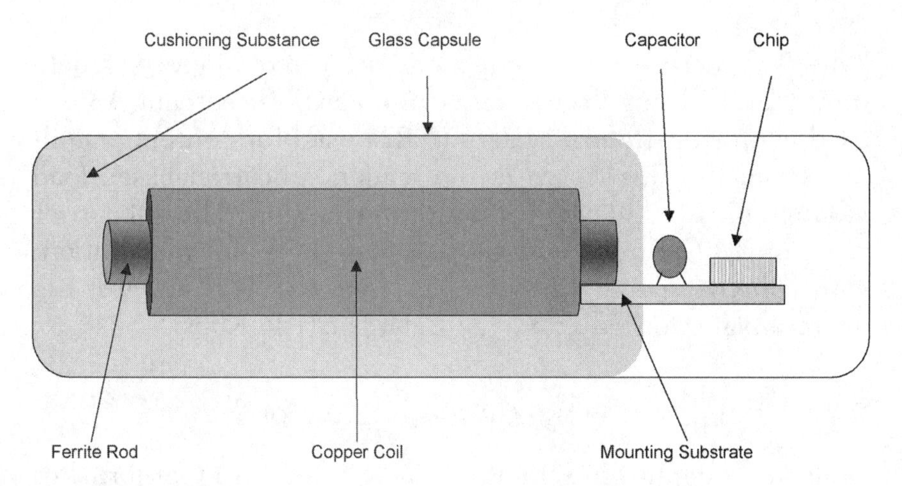

FIGURE 2-30 The anatomy of an RFID transponder; note the large coil at the center of the device.

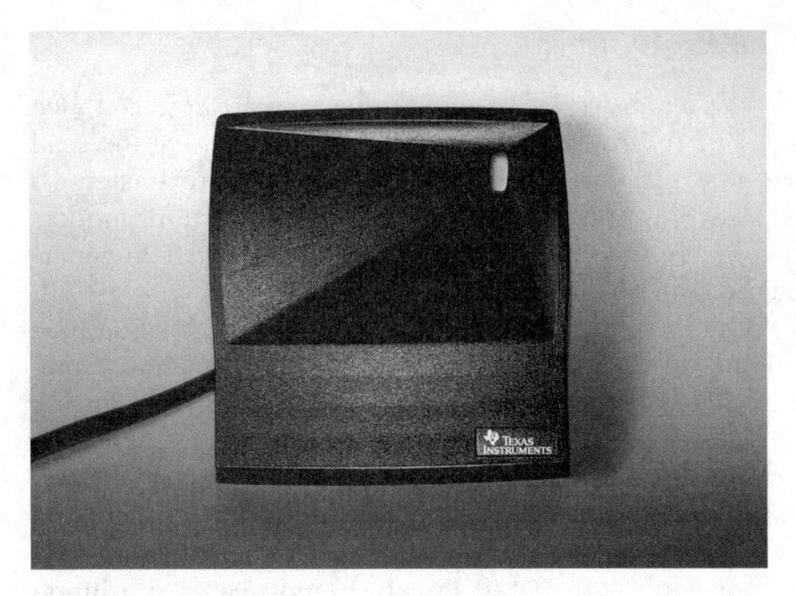

FIGURE 2-31 An RFID reader. *(Photo courtesy Texas Instruments.)*

ISO 14443 Type A

For the engineers in the audience, Type A cards rely on 100 percent *Amplitude Shift Keying* (ASK) as their signal modulation mechanism for card-to-reader transmission. The bidirectional data transmission rate is 106 Kbps.

ISO 14443 Type B

Type B cards use a different modulation technique. For data transmission from the reader to the card, 10-percent ASK is used, with *Non-Return to Zero* (NRZ) used for content encoding. From the smart card to the reader, subcarrier-based load modulation is employed for data transmission. The subcarrier relies on 180-degree *Phase Shift Keying* (PSK) for modulation, with content encoded using NRZ. Like the Type A card, the data transmission rate for Type B cards is 106 Kbps.

Part 3: Initialization and Collision Control

Consider the following situation: A box filled with toothbrushes bound for the ServicePlus warehouse passes beneath an RFID reader in a distributor warehouse on its way to the loading dock. Each toothbrush carries its own RFID transponder, and as the box passes beneath the magnetic field generated by the reader, all of the transponders are energized simultaneously, and all transmit their information simultaneously, causing a traffic jam at the reader. This situation, known as a collision instance, occurs routinely in data communications, and, as a result, protocols must be in place to manage and recover from collisions without experiencing data loss.

Closely related to collision detection and recovery, which will be discussed in more detail in a moment, is interface initialization. When an RFID card enters the field of influence of an RFID reader, the energy field of the reader turns on the transponder tag, causing it to take some predefined action. This process is a carefully orchestrated and highly standardized dance between the two devices. Part 3, therefore, begins with a detailed discussion of the initialization procedure. Of course, initialization and collision control are highly related, because it is quite likely that when a tag is initialized by the presence of the reader's field, the reader is already engaged in communication with another RFID tag that preceded it into the field. Naturally, the differences between Type A and B cards require slightly different control procedures.

Type A Devices

When a Type A device is energized by the field of a reader, it first runs through a sequence of initialization tasks, during which it verifies proper operation and determines (assuming that it's a dual-mode card) whether it is operating in a contact or contactless mode. Once the initialization process completes, the card goes into idle mode, which allows the reader to ignore the card and complete its communication session with another device. While in idle mode, smart cards will not respond to reader commands, thus ensuring that the in-progress communication can complete.

When an idle card receives what is known as a REQUEST-A (REQA) command from a reader, it responds to the interrogating reader with an ANSWER TO REQUEST (ATQA) block.

To guarantee that the tag's ATQA response is not interpreted by another tag as a REQA command from the reader, the data field of the ATQA transmission comprises only 7 bits, as opposed to a normal 8-bit ATQA field sent from the reader. The format of both frames is shown in Figure 2-32.

Once the card has responded to the REQA command, it is placed into ready mode. At this point the reader has identified and responded to the presence of at least one card in its field of influence. It continues by initiating the Dynamic Binary Search Tree Algorithm, a mouthful of capability that translates into the anticollision protocol.

The ladder diagram in Figure 2-33 shows the various stages of card-to-reader communications as the communication link between the two is created.

THE ANTICOLLISION PROTOCOL: HOW IT WORKS

When collisions occur during data transmission, the information sent by the various tags in the reader's field is effectively destroyed due to the fact that it becomes virtually impossible to tell the transmission of one device from that of another. To ensure that an orderly recovery is possible and that each tag gets a chance to transmit in an unimpeded fashion, collision management protocols are used. This methodology is called contention management and is designed to provide a management mechanism for environments in which multiple devices are contending for access to a shared medium. In the case of RFID devices, that shared medium is the attention of the reader.

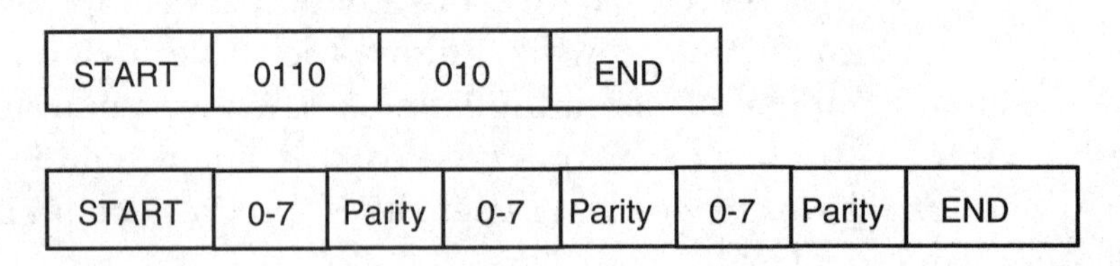

FIGURE 2-32 The reader's REQA frame (above); transponder tag's ATQA (below); note the seven-bit data field in the REQA frame, which uniquely identifies it to prevent confusion.

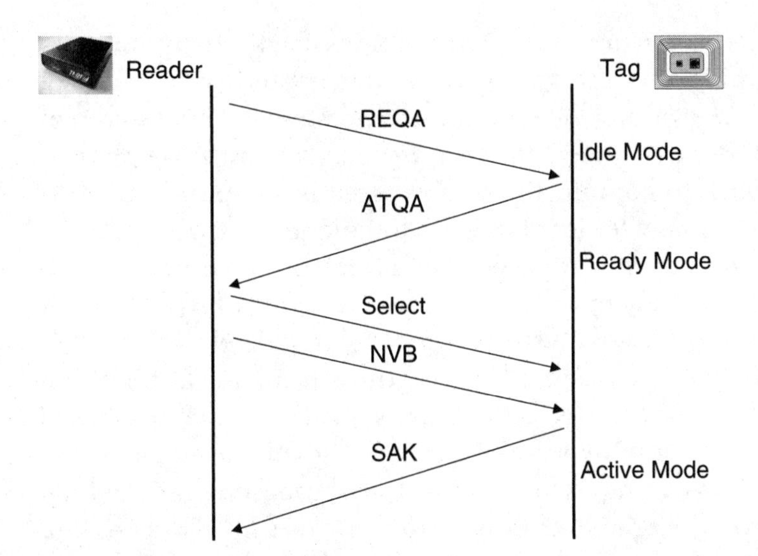

FIGURE 2-33 Ladder diagram showing protocol exchange between a Type A device and the reader as it passes from idle to active mode.

This technique has been around for a long time in data transmission. Perhaps the best-known contention-based medium access scheme is traditional Ethernet, a product developed by 3Com founder and Xerox PARC veteran Bob Metcalfe. In contention-based LANs, devices attached to the network vie for access using the technological equivalent of gladiatorial combat. "If it feels good, do it" is a good way to describe the manner in which they share access (hence the Berkeley Method). If a station wants to transmit, it simply does so, knowing that the possibility exists that the transmitted signal may collide with the signal generated by another station that transmits at the same time. Even though the transmissions are electrical and are occurring on a LAN, there is still some delay between the time that both stations transmit and the time that they both realize that someone else has transmitted. This realization is called a collision, and it results in the destruction of both transmitted messages. In the event that a collision occurs as the result of simultaneous transmission, both stations back off by immediately stopping their transmissions, wait a random amount of time, and try again. This technique has the wonderful name of *Truncated Binary Exponential Backoff*. It's one of those phrases you just *have* to commit to memory because it sounds so good when you casually let it roll off the tongue in conversation.

Ultimately, each station will get a turn to transmit, although how long each may have to wait is based on how busy the LAN is. Contention-based systems are characterized by what is known as *unbounded delay*, because there is no upward limit on how much delay a station can incur as it waits to use the shared medium. As the LAN gets busier and traffic increases, the number of stations vying for access to the shared medium, which only allows a single station at a time to use it, also goes up, which naturally results in more collisions. Collisions translate into wasted bandwidth, so LANs do everything they can to avoid them. We will discuss techniques for this in the contention world a bit later in this chapter.

The protocol that contention-based LANs employ is called *Carrier Sense, Multiple Access with Collision Detection* (CSMA/CD). In CSMA/CD, a station observes the following guidelines when attempting to use the shared network. First, it listens to the shared medium to determine whether it is in use or not—that's the Carrier Sense part of the name. If the LAN is available (not in use), it begins to transmit but continues to listen while it is transmitting, knowing that another station could also choose to transmit at the same time—that's the Multiple Access part. In the event that a collision is detected, usually indicated by a dramatic increase in the signal power measured on the shared LAN, both stations back off and try again—that's the Collision Detection part.

Earlier, we discussed the use of NRZ coding for data transmission. Another coding scheme, known as *Manchester Encoding*, can also be used. In NRZ coding, the value of each individual bit is measured by the voltage level on the transmission channel during any given bit time (shown as t_0 and t_1 in Figure 2-34). For the purposes of this discussion, a binary 0 is

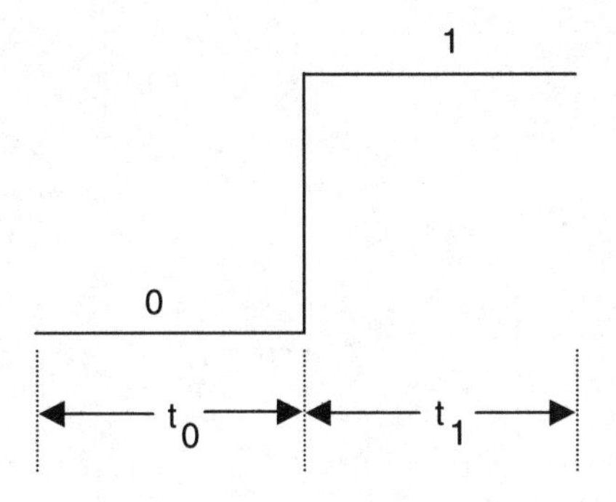

FIGURE 2-34 Non-Return to Zero encoding. The value of each individual bit is measured by the voltage level on the transmission channel at any given bit-time (shown here as t_0 and t_1).

encoded as a low voltage level, and a binary 1 is represented as the presence of a higher voltage.

Assume that one or more of the in-range transponders sends a subcarrier signal to the reader. The reader receives the transmitted signal and interprets it as a high value, which translates into a binary 1. Unfortunately, using NRZ, it is impossible in this situation for the reader to determine whether a received bit stream comes from a single transponder tag or from the combined transmissions of multiple tags. In collision detection scenarios, therefore, NRZ is an inferior encoding scheme.

Consider the example shown in Figure 2-35. Transponder 1 sends the NRZ-encoded bit stream of 010010110 . . . , while transponder 2 transmits 001100100 . . . . The signal received by the reader is the logical combination of the two, 011110110 . . . , which in no way resembles either of the original transmissions. In this situation, an undetectable and therefore unrecoverable collision has occurred. Relate this now to an earlier example, where a box of tagged toothbrushes is making its way through a warehouse on the way to a distributor. In this case, the product would evade detection by the RFID reader! Clearly something better is required; that *something* is Manchester encoding.

MANCHESTER ENCODING

Unlike NRZ Coding where the value of a bit is defined as the steady state value at any specific bit time, Manchester defines

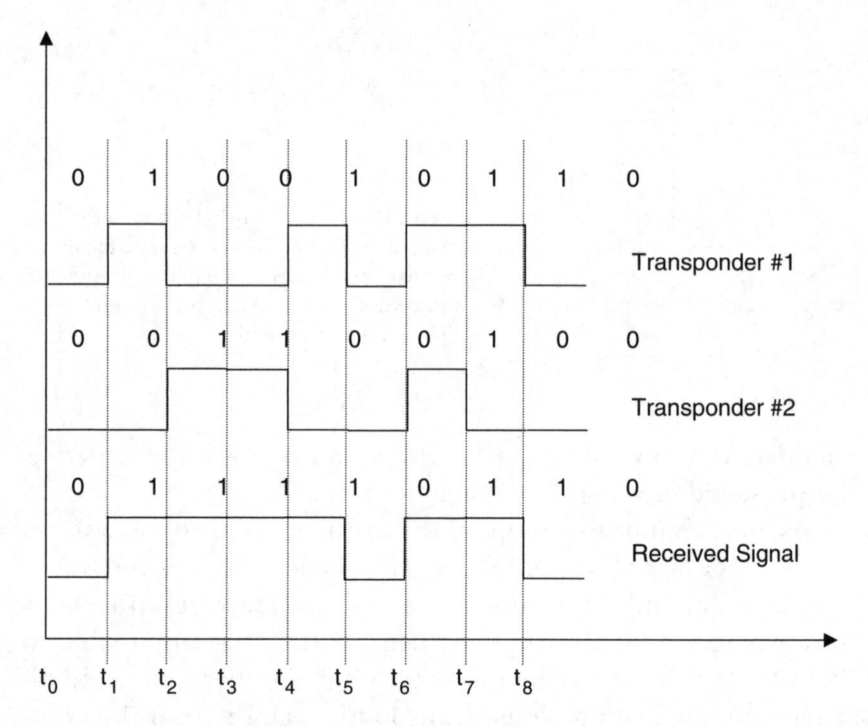

FIGURE 2-35 When NRZ is used in collision-sensitive environments, it becomes difficult to isolate one tag's transmitted signal from that of another.

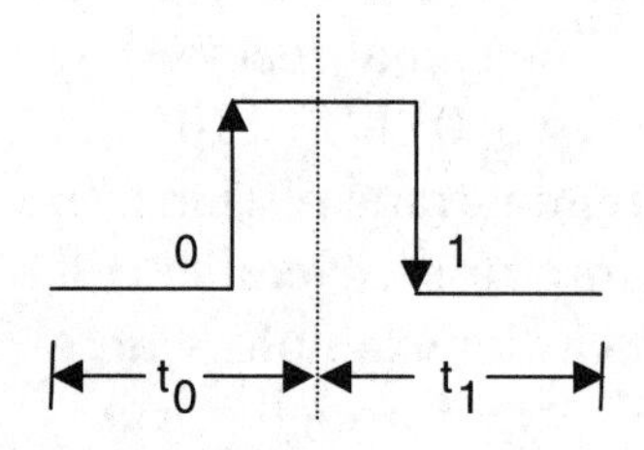

FIGURE 2-36 A Manchester-encoded signal, where a positive transition represents a 0, while a negative transition represents a 1.

the value of a bit by a *transition* from one state to another. In the example shown in Figure 2-36, a 0 is represented as a positive (upward) transition, and a 1 is encoded as a negative (downward) transition. In Manchester, transitions must *always* occur, even if the bit value stays the same. Consider the transmission sequence shown in Figure 2-37. In this example, the

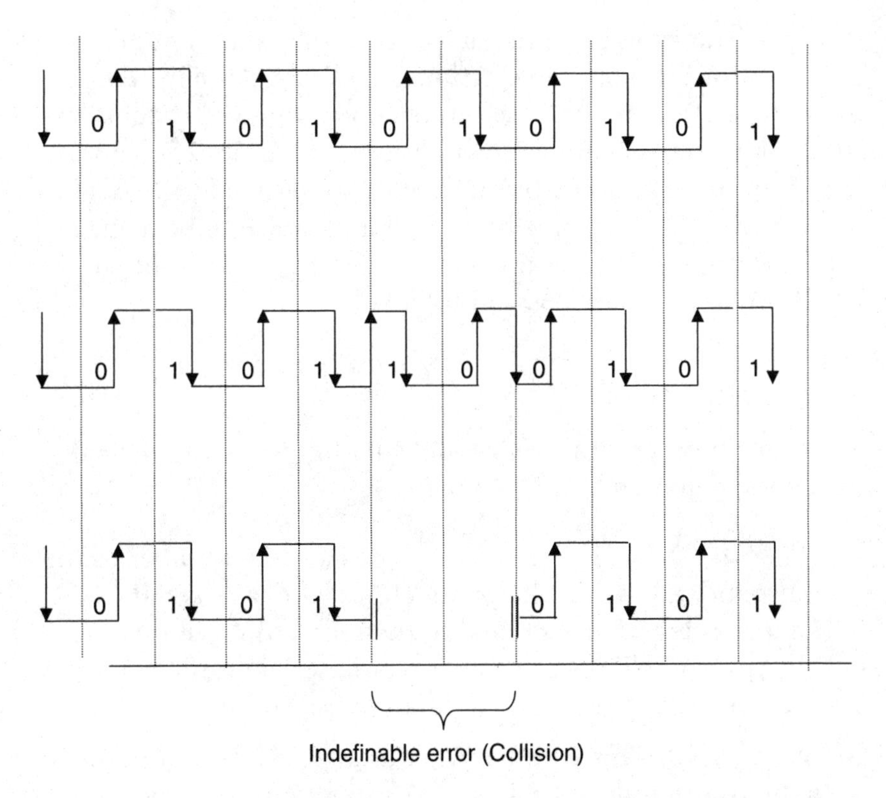

FIGURE 2-37 Because of the Manchester system's inability to discern one transmission from another (by design), it detects a collision at the location shown.

reader detects the presence of a collision, because the logical result of the simultaneously transmitted signal results in a steady state period, which is meaningless in Manchester: Remember, transitions must *always* occur in Manchester-encoded signals. This results in the detection of a collision.

Now that we are all honorary electrical engineers and can discuss encoding schemes with aplomb, let's continue with our discussion.

Binary search techniques rely on a predefined, known sequence of command and response pairs between the transponders that are contending for the attention of the reader and the reader itself. There is a certain amount of gladiatorial combat at work here as they vie for primacy.

Each transponder must have a unique serial number to compete for consideration within a collision domain. For this example, we assume that the transponders in our system have 8-bit addresses, which means that as many as 2^8 (256) unique tags can coexist within the field of a single reader. In reality, most tags have as least 4-byte (32-bit) serial numbers. In our 8-bit example, the addresses will range from binary 0 (00000000) to binary 255 (11111111).

TRANSPONDER SELECTION

The selection process relies on the following four reader-generated commands:

REQUEST_SNR: This command sends a serial number to the transponder as a specific parameter. If the tag's serial number is less than or equal to the received serial number, the tag responds with its own specific serial number. This serves to reduce the field of potential responses.

SELECT_SNR: This command sends a specific serial number to the transponder as a specific parameter. If the tag's serial number is the same as the transmitted address or serial number, it goes into a READY state for receipt and processing of further commands. The tag is now specifically selected. Other tags in the field will only respond to a REQUEST_SNR command, thus eliminating contention on a tag-by-tag basis.

READ_DATA: This command causes the selected tag to transmit its data to the reader.

UNSELECT: This reader-generated command cancels the ongoing session with a particular transponder. The tag is placed back into idle mode and does not respond to any commands, including REQUEST_SNR. This ensures that other tags in the field have higher response priority once a tag has had a turn to transmit. The tag must be removed from the reader's field (and therefore powered down) to be reset for further read-write activities.

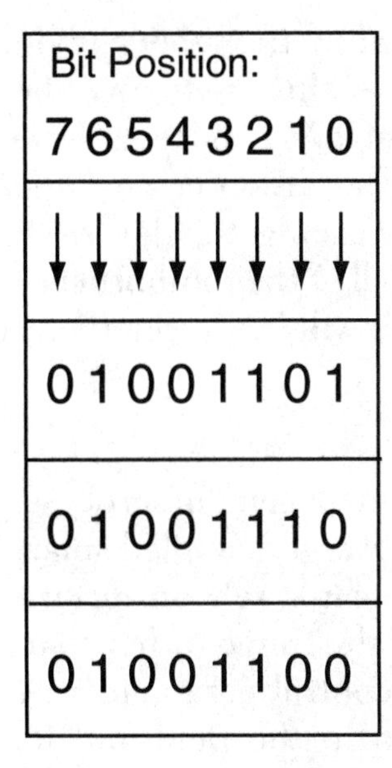

FIGURE 2-38 Serial number transmission by three tags in response to reader-transmitted REQUEST_SNR frame.

COLLISION MANAGEMENT IN ACTION

In step one of our example, the reader broadcasts the REQUEST_SNR command, where SNR=11111111 (the highest possible value). As a result, all of the tags in the field will respond, and all will respond in a synchronized fashion, which is required for the collision detection mechanism to work properly.

Let's assume that three transponders are within range of the reader when it broadcasts the REQUEST_SNR command. Simultaneously, they respond by transmitting their serial numbers, which in this case are pictured in Figure 2-38.

As a result of the Manchester encoding used in the transmission interface between the cards and the reader, the reader can identify collision points and can therefore "cull the herd" of offending devices. In this case, collisions (superimpositions of dissimilar bit values) occur in bit positions zero and one. Bit two is the highest bit in the sequence where a collision took place. This leads the reader to conclude that at least two tags remain. As the collisions occur, the reader "prunes the tree," using the commands described earlier. As one tag succeeds and is allowed to transmit, it is subsequently deactivated by the reader upon completion of its transmission sequence; thus reducing the number of active tags "cluttering the field." This technique allows the reader to elegantly manage the presence of large numbers of tags while ensuring that they each get a chance to be interrogated by the reader. As a result, all of the toothbrushes in the box on the way to the loading dock will be detected and registered.

BACK TO TYPE A CARDS

Now that we understand the collision management process, let's return to our discussion of Type A proximity-coupled smart cards. The procedures that govern these cards rely on bit-oriented protocols, meaning that each bit in the frame (other than the actual payload) can convey specific control parameters by virtue of nothing more than its position in the field and its value. Because the interface between the reader and the transponder tag is bidirectional, the frame transmit direction needs to be periodically reversed. The reader uses a search criterion to identify the desired tag serial number. The reader then uses the *NUMBER OF VALID BITS* (NVB) field to identify the length of the desired search parameter.

Once a card has been selected via the serial number elimination process of collision management, the reader transmits the entire selected serial number using the SELECT command. The selected card acknowledges by responding with a SAK message, which results in the tag being placed into active mode by the reader.

The nice thing about standards is the fact that there are so many to choose from. Although most tags have a 4-byte serial number, the field size is far from universal. Some cards have 7-byte serial numbers; others have 10. If the card has one of the larger serial number fields, this will be indicated by the presence of SET cascade bit in the SAK response. Of course, these larger serial numbers require the collision mechanism to be invoked multiple times during the weeding out process, but it does so flawlessly.

TYPE B CARDS

Type B cards are not all that different from their Type A counterparts, with a few notable exceptions. When a Type B card is energized by a reader, it goes through startup and diagnostic routines, after which it goes into idle mode while it waits for an REQB command. The REQB command kicks off the collision management routine in Type B devices, which is a Slotted Aloha collision scheme, explained in the following section. The number of available slots is determined by the reader and communicated through the REQB broadcast. It must also be noted here that the REQB protocol identifies an *Application Family Identifier* (AFI) field that provides for granular searching at the application level. Some of these applications are shown in Table 2-1.

Once a transponder tag receives an REQB command, it verifies the presence of the AFI-identified application onboard. If present, it verifies the number of available anticollision slots (M). If M is greater than one, a random number generator onboard the tag selects a slot number that the tag will use for transmission to the reader. To ensure synchronization between cards and slots, the reader transmits its own slot identifier as a synchronization signal. The tag waits until the slot marker of the selected slot is received (READY REQUESTED STATE) and then responds to the REQB broadcast with an ANSWER TO REQUEST B (ATQB).

THE REQB FRAME

An REQB frame is shown in Figure 2-39. The first field on the left, Apf, is the Anticollision Prefix. Apf is assigned a unique value that cannot (by design) appear in the *Node Address* (NAD) field. The *Application Family Identifier* (AFI) identifies the selected application.

The Parameter field can indicate a variety of things including the number of available timeslots in the Slotted Aloha environment, the maximum allowable frame size, and the maximum

TABLE 2-1. RFID Application Groups and Subgroups

Application	Description of Sample Applications	AFI Bits 7–4 Application Group	AFI Bits 3–0 Subgroup
All application groups and subgroups	All application groups and subgroups	0000	0000
All subgroups of an application group	All subgroups of an application group	—	0000
Only subgroup Y of application X	Only subgroup Y of application group X	X	Y
Transport	Airlines, taxis, local transport, regional trains, etc.	0001	—
Payments	Banks, ticket purchases, electronic bill payment, funds transfer	0010	—
Identification	Passport, driver's license, corporate identity card	0011	—
Telecommunications	GSM SIM card, telephone third-party billing card	0100	—
Medicine	Health insurance identification, medical history, drug allergies	0101	—
Multimedia	Video-on-Demand, Pay TV, Internet access	0110	—
Games	Lottery card, casino billing card	0111	—
Data storage	File transfer	1000	—
Reserved	Reserved for future applications	1001–1111	—

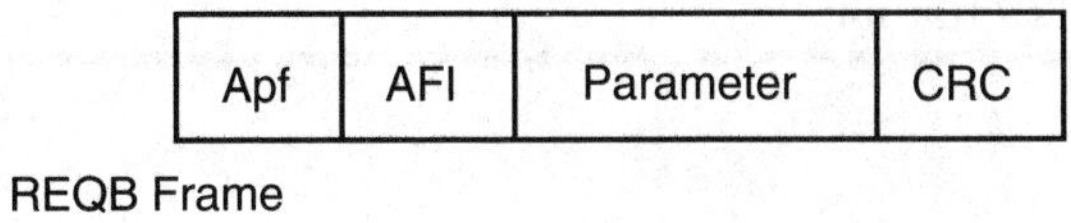

Apf	AFI	Parameter	CRC

REQB Frame

FIGURE 2-39 The REQB frame, used to initiate communication between a reader and a specific transponder tag.

transmission rate allowable. The frame size is directly related to the amount of memory onboard the card; small cards with limited RAM may require smaller frame sizes.

For setting the number of available slots, the *M* parameter is set in the Parameter field according to Table 2-2. Finally, the 2-byte *Cyclic Redundancy Check* (CRC) field provides for error correction in the transmitted frame to ensure error detection and correction.

For example, a reader might send an REQB frame with the following parameters: Find a valid ISO 14443 Type B device with the COM port set to 9,600 bps, no parity, and 1 stop bit (8N1), specified with a single slot. Then, validate the data that is received in return. Transponders that are in the transmission field should respond with a valid ATQB frame.

THE ATQB FRAME

The ATQB frame, transmitted by the transponder tag in response to the successful receipt of a valid REQB frame, comprises five fields. The first, shown in Figure 2-40, is the Apa field (byte 1), which uniquely identifies the frame, similar to the Apf field in the REQB transmission. Next is the PUPI field (bytes 2 through 5), which contains the unique serial number of the transponder tag that is responding, followed by the Application Data field (bytes 6 through 9), the Protocol Information field (bytes 10 through 12), and a CRC (bytes 13 through 14). Once the reader has received a valid ATQB frame from one or more cards in the array, a card can be selected. This is done through the transmission of an initial application command transmitted from the reader to a tag. The frame's structure, shown in Figure 2-41, is

TABLE 2-2 M Parameter Definitions

M PARAMETER (BITS 0–2)	NUMBER OF SLOTS (N)
000	1
001	2
010	4
011	8
100	16
101	Reserved for future use
11x	Reserved for future use

Apa	PUPI	Application Data	Protocol Info	CRC

FIGURE 2-40 The ATQB frame.

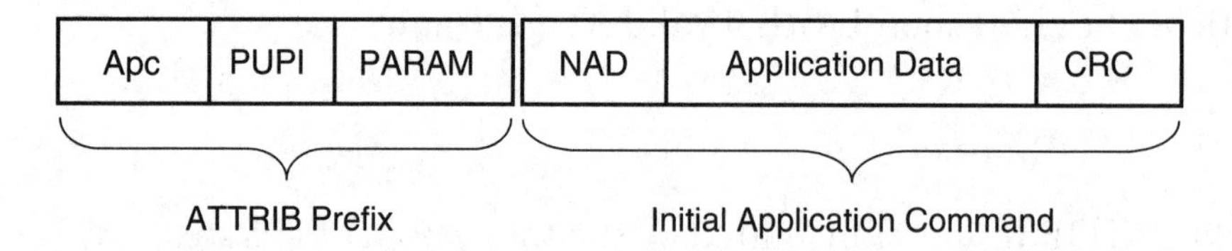

FIGURE 2-41 The application command, used by the reader to select a specific transponder tag through the use of the unique PUPI identifier.

based on a standard frame but also includes an *Attribute* (ATTRIB) prefix. The ATTRIB prefix includes the selected card's unique PUPI, and a parameter byte that details the communications requirements for the RF interface between the card and the reader, such as maximum transmission wait times and the like. Once that agreement has been met, the card can engage whatever application it is designed to execute. Figure 2-42 shows the ladder diagram with the progression from idle to active mode.

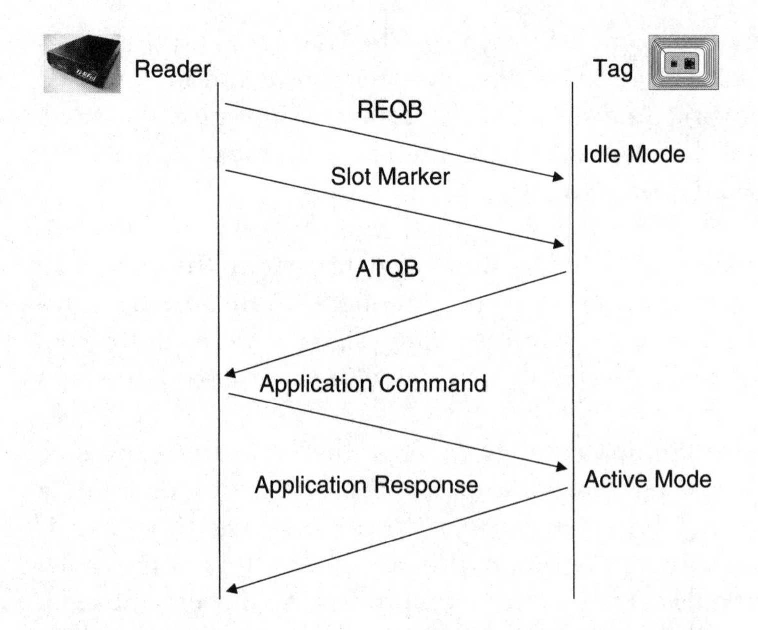

FIGURE 2-42 Ladder diagram showing protocol exchange between a Type B device and the reader as it passes from idle to active mode.

SLOTTED ALOHA: HOW IT WORKS

When LANs began to emerge in the 1970s, a great deal of research on their operation and use was going on at research facilities around the world. One of the most promising of these (besides Xerox PARC) was at the University of Hawaii's Manoa Campus in Honolulu. There, one of the first LANs was created. It was called ALOHANet and was used for interisland data transmission.

Before the arrival of ALOHANet, communications systems shared a number of operational features. Transmitted data was converted to analog using a modem, and transmission was accomplished via the telephone network. The typical connection was point to point and established manually.

ALOHANet, on the other hand, was a truly automated data network. Machines attached to ALOHANet could transmit completely autonomously any time they desired, and there was no real limit on the number of systems that could be attached to the network. And because the transmission medium was a radio signal that operated continuously, the cost to operate the system scaled rather nicely.

ALOHANet was one of the first contention-based networks. It was also one of the first systems to demonstrate the challenge of collision management. If multiple nodes transmitted simultaneously, they collided and both signals were destroyed. Clearly, some form of collision detection and correction was required.

The most common way to manage this problem is to place each transmitting system into its own frequency domain, a technique that has been around for a very long time and is known as *frequency division multiplexing*. This technique, however, is complex, in that the communicating devices must be frequency agile, a capability that was well beyond the capabilities of computers in those early days—at least, at a reasonable price. For ALOHANet, it further required that hundreds, perhaps thousands, of channels be available, and given the fact that radio spectrum was (and is) a limited and precious resource, this model was not considered practical for LANs.

A second possible solution was to have time slots made available into which each node could transmit on a round-robin basis, a technique known as *time division multiplexing* (TDM) and commonly used in carrier systems such as T-carrier. TDM is easy to implement because the transmitting systems share a single radio frequency but do so on a turn-taking basis. On the other hand, if a particular system has nothing to send during its dedicated timeslot, the slot is wasted. Clearly, neither of these was ideal for the purposes of ALOHANet.

Instead, ALOHANet was based on a relatively new technique called *Carrier Sense Multiple Access* (CSMA). In CSMA, there is no multiplexing per se. Instead, each system listens to the shared medium (the radio channel) to determine whether it is idle, and if it is, they begin to transmit.

It would seem that the first node to transmit might be able to keep the channel for as long as it required, effectively isolating all other systems. To avoid this problem, ALOHANet forced transmitting systems to use a very small maximum packet size and to send them one at a time with gaps in between. This allowed other computers to transmit their own packets in between, allowing everyone to share the radio medium and transmit simultaneously.

Naturally, problems began to crop up as soon as the network got busy. If two or more systems tried to broadcast at the same time, collisions occurred. After all, this was a relatively slow system, and there was transmission latency that would often prevent one system from hearing if another machine was "on the line." To deal with this issue, ALOHANet engineers came up with a clever alternative. As soon as a system transmitted a packet, it was required to listen for a retransmission of its own packet returned from a central hub. As soon as they heard the retransmission of their own packet, the system was allowed to transmit another. If, however, the system failed to receive the retransmission—an occurrence that would typically mean that a collision had occurred—the system was instructed to wait a random amount of time and try again. Because each system would typically choose a random time to wait before retransmitting, it follows that one of them would be the first to retry, and that other computers on the network would be able to tell that the channel was in use when they attempted to transmit. Under most circumstances, this avoided collisions, as shown in Figure 2-43.

The downside of CSMA is that when the network gets busy, collisions rise dramatically. In the ALOHANet network, maximum channel utilization hovered around 18 percent, and any attempts to become more efficient were typically met with increased collisions. This would often result in what came to be known as *collision collapse*.

An alternative to traditional CSMA was needed for ALOHANet. The *Slotted Aloha* protocol, illustrated in Figure 2-44, raised the channel utilization level to approximately 35 percent, a significant improvement over traditional CSMA. With Slotted Aloha, a centralized clock transmitted a timing signal to all systems connected to the network. Those systems were only

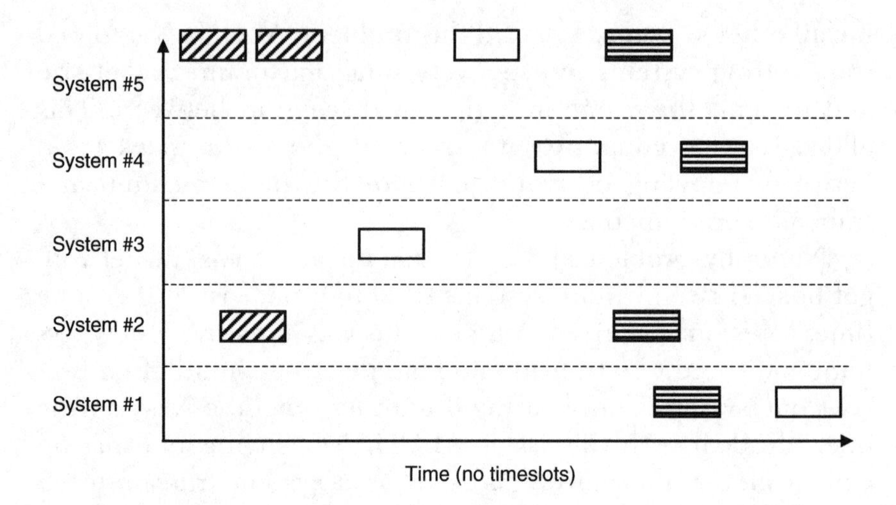

FIGURE 2-43 Because there are no dedicated timeslots or frequency allocations in traditional CSMA, transmitted packets collide with one another. This was the case in the early ALOHANet system.

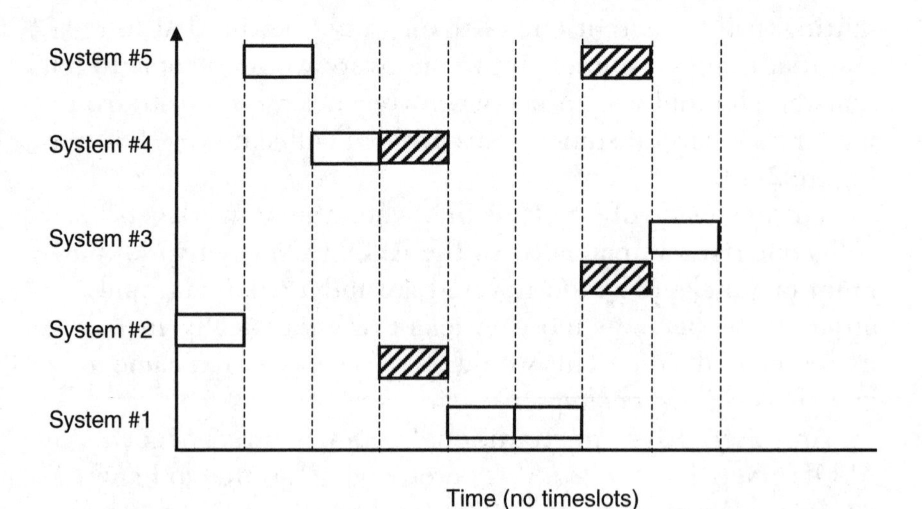

FIGURE 2-44 In Slotted Aloha systems, the provisioning and use of predetermined timeslots guarantees that collisions, while not eliminated, are dramatically reduced.

allowed to transmit immediately upon receiving a clock signal. In the unlikely scenario of a single device network, Slotted Aloha guaranteed that a collision would never take place. On the other hand, if multiple stations had packets to transmit, Slotted Aloha guaranteed a collision, resulting in the functional loss of that timeslot. The result is a reduction in collision counts of nearly 50 percent.

The ALOHANet itself used 9,600 bps modems. It utilized two 100 kHz channels, one a *broadcast channel* at 413.475 MHz, the other a *random access channel* that operated at 407.350 MHz. The network was deployed in a star configuration, with a central computer receiving messages on the random access channel and rebroadcasting them to all nodes on the broadcast channel. This reduced the number of collisions possible. Later changes added signal repeaters that also served as hubs, greatly increasing the reliability and size of the network. Send and receive packets had 32-bit headers with 16-bit parity checks, followed by as many as 80 bytes of data and a 16-bit parity check.

THE AIR INTERFACE: DATA COMMUNICATIONS PROTOCOLS

Once a conversation is underway between a card and a reader, the reader can begin to transmit commands to the card: read, write, and data process, as appropriate. The transmission protocols portion of the ISO 14443 standard defines the functions that typically take place at OSI Layer Two: error detection and correction. Type A cards, which are somewhat variable in their operation, use this time to transfer data transmission rates, frame sizes, and other parameters that are unique to specific tags or readers. Type B cards, on the other hand, do the same thing during the data exchange process that occurs as part of the anticollision function. As a consequence, the data exchange protocol can begin immediately. In Type A devices, the protocol must first be activated. When a Type A card is selected during the initial collision control process, it confirms its selection by transmitting an SAK response to the reader. (See Figure 2-45.) The SAK frame includes an indicator that tells whether the

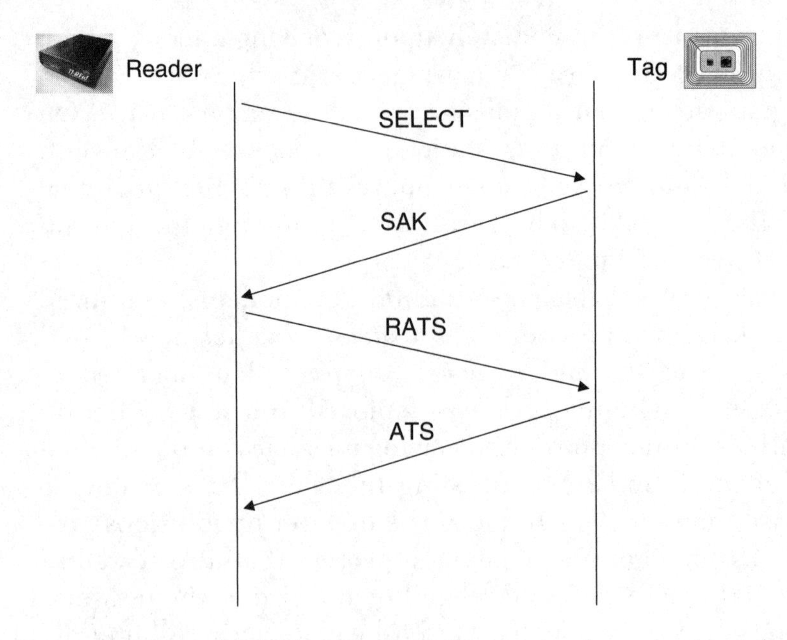

FIGURE 2-45 Ladder diagram showing protocol exchange between a Type A device and the reader as the reader initiates the data transfer process.

operational protocol embedded in the card is in compliance with ISO 14443–4, or is in fact proprietary. If the protocol is ISO 14443–4 compliant, the reader transmits a *REQUEST FOR ANSWER TO SELECT* (RATS) command to the card, which causes the card to respond with an *ANSWER TO SELECT* (ATS). The RATS command includes a *Frame Size Device Integer* (FSDI) field as well as a *Card Identifier* (CID) field. The FSDI field specifies the maximum number of data bytes that can logically be transmitted from the tag to the reader in a single block. The CID field contains a unique identifier that allows a reader to maintain open conversations with several cards simultaneously, querying each individually through the unique CID identifier.

The RATS command results in the transmission of the ATS, which is analogous to the ATQA response described earlier for use with close-coupling smart cards. It provides for discrete control of the interaction between the card and the reader based on the functional limitations of the card's operating system. The ATS can in fact include a considerable amount of additional optional information, including the following:

Frame Size Card Integer (FSCI): This field identifies the maximum number of bytes that can be transmitted by a card to a reader in a single block. Valid numbers are 16, 32, 64, etc., up to 256.

Data Rate Send (DR): This field identifies valid transmission speeds from the card to the reader during data transmission. Supported values include 106 kbps, 204 kbps, 408 kbps, and 816 kbps.

Data Rate Send (DS): This field identifies valid transmission speeds from the card to the reader during data transmission. Supported values include 106 kbps, 204 kbps, 408 kbps, and 816 kbps.

Frame Waiting Integer (FWI): This value defines the maximum time that a reader must wait for a response from a card. If the time expires and no response is forthcoming, a timeout occurs and recovery procedures go into effect.

Startup Frame Guard Integer (SFGI): This is, in effect, a warmup procedure. Once the ATS has been sent to the reader followed by the first application command from the reader back to the card, the SFGI defines how long the reader must wait for a response to that initial application command.

NAD, CID Supported: Because of variable limitations in the capabilities of different cards' operating systems, these fields stipulate whether the card in question supports the NAD and CID parameters.

Historical Bytes: This is a user-definable field that can be used to carry system or card-specific data such as application or operating system version numbers. Like all of these paramaters, it is optional.

INITIATING DATA TRANSMISSION

As soon as the reader receives the ATS response from the selected card, it begins the data exchange process. The initial transmission rate, described earlier, is 106 kbps, but at this point the reader has the option to modify the rate in either direction by issuing a *Protocol Parameter Selection* (PPS) broadcast, which allows the reader to increase the rate, assuming the reader received information from the card (in the ATS, remember?) in the DR or DS fields, which indicated the card's ability to support higher transmission rates.

APPLICATION PROTOCOL SUPPORT

To understand the data transmission interface between the card and the reader, it is important to understand the nature of data communications protocols. And the best way to do that is to understand the inner workings of the OSI Model. Readers familiar with data transmission and layered protocols may want to skip this section; for others, please read on.

UNDERSTANDING OSI

For our purposes, it is good to understand the fundamental model around which all data communications environments are built, including the environment within which RFID operates. Perhaps the best-known protocol model is the *Open Systems Interconnection* (OSI) Reference Model, most often referred to as the OSI Model. Shown in Figure 2-46 and consisting of seven layers, it provides a logical way to present data communications and is based on the following simple rules. First, each of the seven layers must perform a clearly defined set of tasks, which are unique to that layer. Second, each layer depends upon the services of the layers above and below it to do its own job. Third, the layers have no idea how the layers around them do what they do; they simply know that they do it. This is called transparency. Finally, there is nothing magic about the number *seven*. If the industry should decide that an eighth layer is needed on the model, or that layer six is no longer needed, then the model will be changed. The key is functionality.

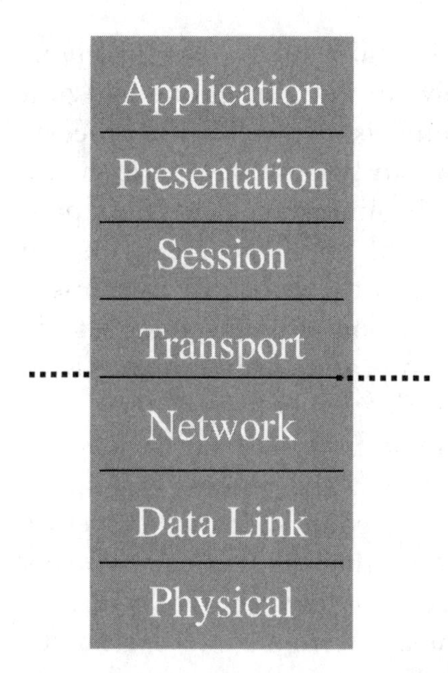

FIGURE 2-46 The seven-layer OSI Reference Model.

It is also important to understand that the OSI Model is nothing more than a conceptual way of thinking about data communications. It isn't hardware or software; it merely simplifies and groups the processes of data transmission so that they can be easily understood and manipulated, and so that complex systems can be built around them with minimal difficulty.

The functions of the model can be broken into two pieces, as illustrated by the dashed line between layers three and four that divides the model into the *chained layers* (below the line) and the *end-to-end layers* (above the line).

The chained layers include layers one through three: the Physical Layer, the Data Link Layer, and the Network Layer. They are responsible for providing a service called *connectivity.* The end-to-end layers on the other hand include the Transport Layer, the Session Layer, the Presentation Layer, and the Application Layer. They provide a service called *interoperability.* The difference between the two services is important.

Connectivity is the process of establishing a physical connection so that electrons can flow correctly from one end of a circuit to the other. Little intelligence is involved in the process; it occurs, after all, pretty far down in the protocol ooze of the OSI Model. Connectivity is critically important to network people because it represents their lifeblood. Customers, on the other hand, are typically only aware of the criticality of connectivity when it isn't there for some reason. No dial tone? Visible connectivity. Can't connect to the Internet? Visible connectivity. Dropped call on a cell phone? Visible connectivity.

Interoperability, however, is something that customers are much more aware of. Interoperability is the process of guaranteeing *logical connectivity* between two communicating processes over a physical network. It's wonderful that the lower three layers give a user the ability to spit bits back and forth across a wide area network. But what do the bits mean? Without interoperability, that question cannot be answered. For example, the e-mail application that runs on a PC and the e-mail application that runs on a mainframe are logically incompatible with each other for any number of reasons. They can certainly swap bits back and forth, but without some form of protocol intervention, the bits are meaningless. If the PC creates an e-mail message that is compressed, encrypted, ASCII encoded, and shipped across logical channel 17, do the intermediate switches that create the path over which the message is transmitted care? Of course not. Only the transmitter and receiver of the message that house the applications that will have to interpret the message care about such things. The intermediate switches care that they have electrical connectivity, that they can see the bits, that they can determine whether they are the *right* bits, and whether they are the intended recipient or not. Therefore, the end devices, the sources and sinks of the message, must implement all seven layers of the OSI Model, because they must not concern themselves only with connectivity issues, but also with issues of interoperability. The intermediate devices, however, only care about the functions and responsibilities provided by the lower three layers. Interoperability, because it only has significance in the end devices, is provided by the end-to-end layers—Layers Four through Seven.

Connectivity, on the other hand, is provided by the chained layers, layers one through three, because those functions are required in every link of the network chain—hence the name.

LAYER BY LAYER

The OSI Model relies on a process called *enveloping* to perform its tasks. If we return to our earlier e-mail example, we find that each time a layer invokes a particular protocol, it wraps the user's data in an envelope of overhead information that tells the receiving device about the protocol used. For example, if a layer uses a particular compression technique to reduce the size of a transmitted file, and a specific encryption algorithm to disguise the content of the file, then it is important that the receiving device be made aware of the techniques employed so that it knows how to decompress and decrypt the file when it receives it. As a result, a significant amount of overhead must be transmitted with each piece of user data. The overhead is needed, however, if the transmission is to work properly. So as the user's data passes down the stack from layer to layer, additional information is added at each step of the way. In practice, the e-mail message to be transported is handed to layer seven, which performs Application-Layer functions and then attaches a header to the beginning of the message that explains the functions performed by that layer so that the receiver can interpret the message correctly. When the receiving device is finally handed the message at the Physical Layer, each succeeding layer must open its own envelope until the kernel—message—is exposed for the receiving application. Thus OSI protocols work like the layers of an onion. After peeling back layer after layer, the core message is exposed.

LAYER SEVEN: THE APPLICATION LAYER

The network user's application (Eudora, Outlook, Outlook Express, PROFS, etc.) passes data down to the uppermost layer of the OSI Model, called the Application Layer. The Application

Layer provides a set of highly specific services to the application that have to do with the *meaning* or *semantic content* of the data. These services include file transfer, remote file access, terminal emulation, network management, mail services, and data interoperability. This interoperability is what allows our PC user and our mainframe-based user to communicate: The Application Layer converts the application-specific information into a common, *canonical* form that can be understood by both systems. A canonical form is a form that can be understood universally. The word comes from *canon,* which refers to the body of officially established rules or laws that govern the practices of a church. The word also means an accepted set of principles of behavior that all parties in a social or functional grouping agree to abide by.

LAYER SIX: THE PRESENTATION LAYER

The Presentation Layer provides a more general set of services than does the Application Layer, and they have to do with the structural *form* or *syntax* of the data. These include *code conversion* (ASCII to EBCDIC, for example); *compression*, using such services as the *Moving Picture Experts Group* (MPEG) or the *Joint Photographic Experts Group* (JPEG); and *encryption*, including *Pretty Good Privacy* (PGP), *Public Key Infrastructure* (PKI), and so on. Note that these services can be used on any form of data: Spreadsheets, word processing documents, and rock music can all be compressed and encrypted.

LAYER FIVE: THE SESSION LAYER

The Session Layer ensures that a logical relationship is created between the transmitting and receiving applications. It guarantees, for example, that a PC user in Singapore receives his or her e-mail and *only* his or her e-mail. This requires the creation and assignment of a logical session identifier.

Layer Five also shares responsibility for security with the Presentation Layer. You may have noticed that when you log in to your e-mail application, the first thing the system does is ask for a login ID, which you dutifully enter. The ID appears in the appropriate field on the screen. When the system asks for your password, however, the password does not appear on the screen: The field remains blank or is filled with stars. This is because the Session Layer knows that the information should not be displayed. When it receives the correct login ID, it sends a command to the terminal (your PC), asking you to enter your password. It then immediately sends a second message to the terminal telling it to turn off local echo so that your keystrokes are not echoed back on to the screen. As soon as the password has been transmitted, the Session Layer issues a command to turn local echo back on again, allowing you to once again see what you type.

Layer Four: The Transport Layer

After adding a header, Layer Five hands the steadily growing *Protocol Data Unit*, or PDU, down to the Transport Layer. This is the point where we first enter the network. Until now, all functions have been software-based, and in many cases a function of the operating system.

The Transport Layer's job is simple: to guarantee end-to-end, error-free delivery of the entire transmitted message, not bits, not frames or cells, not packets, but the entire message. It does this by taking into account the nature and robustness of the underlying physical network over which the message is being transmitted, including the following characteristics:

- Class of service required
- Data transfer requirements
- User interface characteristics
- Connection management requirements
- Specific security concerns
- Network management and reporting status data

LAYER THREE: THE NETWORK LAYER

We have now left the end-to-end layers, which govern *interoperability*, and entered the realm of the chained layers, which govern *connectivity*. The Network Layer, the uppermost of the three chained layers, has two key responsibilities: routing and congestion control. Routing is the process of intelligently selecting the most appropriate route through the network for the packets, and congestion control is the process that ensures that the packets are minimally delayed (or at least equally delayed) as they make their way from the source to the destination.

LAYER TWO: THE DATA LINK LAYER

The Data Link Layer is responsible for ensuring bit-level integrity of the data being transmitted. In short, its job is to make the layers above believe that the world is an error-free and perfect place. When a packet is handed down to the Data Link Layer from the Network Layer, it wraps the packet in a *frame*. In fact, the Data Link Layer is sometimes called the *frame layer*. The frame built by the Data Link Layer is composed of several fields, shown graphically in Figure 2-47, that give the network devices the ability to ensure bit-level integrity and proper delivery of the packet, now encased in a frame, *from switch to switch*. Please note that this is different from the Network Layer, which concerns itself with routing packets to the final destination. Even the addressing is unique: Packets contain the address of the ultimate destination, which is used by the network to route the packet properly; frames contain the address of the next link in the network chain (the next switch), used by the network to move the packet along, switch by switch.

PCB	CID	NAD	Application Data	CRC

FIGURE 2-47 The ISO 14443 data frame, showing the higher layer user data (Application Data) transported within the Layer Two Frame.

LAYER ONE: THE PHYSICAL LAYER

Once the CRC is calculated and the frame is fully constructed, the Data Link Layer passes the frame down to the Physical Layer, the lowest layer in the networking food chain. This is the layer responsible for the physical transmission of bits, which it accomplishes in a wide variety of ways. The Physical Layer's job is to transmit the bits, which include the proper representation of 0a and 1s, transmission speeds, and physical connector rules. For example, if the network is electrical, then what is the proper range of transmitted voltages required to identify whether the received entity is a 0 or a 1? Is a 1 in an optical network represented as the presence of light or the absence of light? Is a 1 represented in a copper-based system as a positive or as a negative voltage, or both? Also, where is information transmitted and received? For example, if pin two is identified as the transmit lead in a cable, what lead is data received over? All of these physical parameters are designed to ensure that the individual bits are able to maintain their integrity and be recognized by the receiving equipment.

OSI SUMMARY

We have now discussed the functions carried out at each layer of the OSI Model. Layers Six and Seven ensure application integrity; Layer Five ensures security; and Layer Four guarantees the integrity of the transmitted message. Layer Three ensures network integrity; Layer Two, data integrity; and Layer One, the integrity of the bits themselves. Thus, transmission is guaranteed on an end-to-end basis through a series of protocols that are interdependent and which work closely to ensure integrity at every possible level of the transmission hierarchy.

THE OSI MODEL AND THE RFID INTERFACE

The process of initiating a physical and logical linkage between a transponder tag and a reader, the exercise of guaranteeing that data is transmitted flawlessly, and the initialization of card-based applications that return data to the reader are all defined under the tenets of the OSI Model's protocols. The Physical Layer, which manages the physical interface between the tag and the reader (in this case an air interface) defines such parameters as bit rate, electrical signal representation (the ASK, NRZ, and Manchester signaling, for example), and other transmission characteristics. The functions of this layer are defined in ISO 14443–2; there are two standardized procedures.

Layer Two, the Data Link Layer, is responsible for a variety of functions including error detection and correction, framing, transparency, link control, and point-to-point addressing. In RFID, the Data Link Layer is responsible for address management (using the CID field), transmission of sequential data blocks, link control such as timeout management and collisions as appropriate, and of course, error management.

Layers Three and Four are not really used in RFID implementations, for reasons that should be clear to the reader. The Network Layer, Layer Three, which manages routing and congestion control, is not required because all links in RFID systems are point to point (there are no intermediate switches) and therefore require no routing function. Congestion control is also a nonissue, because in most cases there is a one-to-one communications relationship between the reader and a particular card.

Layer Four, the Transport Layer, ensures the end-to-end transmission of a complete message. Because there is no complex network routing involved in RFID systems, there is no real need to perform complex Transport Layer functions. Those that are required are performed at a more rudimentary level. Layer Five, the Session Layer, is not required, but the security capabilities of Layer Six, the Presentation Layer, may be required for encryption implementations. Although in these kinds of systems encryption is often accomplished as part of a hard-coded function embedded on the chip, performed at the Data Link Layer.

Layer Seven is very much required. The applications that are executed onboard the chip in the transponder must be able to send and receive information to and from the reader as required.

The layers of OSI used by RFID and their associated standards are shown in the following table.

OSI Layers	Associated Standards
Layers 6 and 7 (Presentation, Application)	ISO 7816–4 ISO 7816–7 Various vendor-specific proprietary protocols
Layer 2 (Data Link)	ISO 14443–4 ISO 7816–3 for contact card implementations
Layer 1 (Physical)	ISO 14443–2 for Type A, B devices ISO 7816–2 for contact card implementations

Let's take a quick look at the data transmission process in light of our discussion of layered protocols. Once the transponder has been activated by the ATS command, it goes into a wait state as it anticipates further commands from the reader. It is important to note that the relationship between transponders and readers is the same as the relationship between mainframes and terminals: The mainframe is the primary device (*the server),* whereas the terminal is the secondary device (*the client*). Interactions between the mainframe and the terminal are *always* initiated by the reader, and commands are *always* transmitted by the reader. The transponder's job is to respond, period.

The basic structure of the Layer Two frame used in RFID data exchange is shown in Figure 2-47. The frame consists of a *Protocol Control Byte* (PCB), which is used to identify the type of data block that is being sent. There are three types of data blocks: the *Supervisory* or *S-Block,* which manages higher-layer protocols; the *Recovery* or *R-Block,* which identifies and manages recovery from transmission errors; and the *Information* or

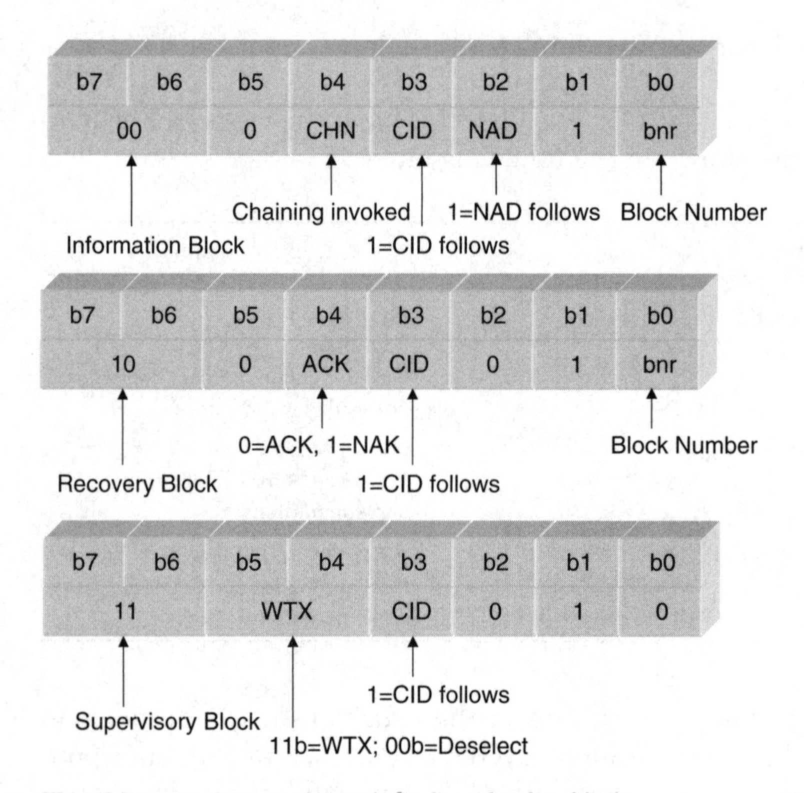

FIGURE 2-48 Frame structure of I, R, and S data blocks.

I-Block, which identifies blocks carrying user data. The structure of each block is shown in Figure 2-48.

Following the PCB field is the optional CID field, which was discussed earlier. You will recall that the CID field is used by the reader to select a specific tag from among a collection of active devices. The NAD field, also optional, is a node address identifier and is used to ensure protocol compatibility between the two ISO standards 7816–3 (Information technology, identification cards, integrated circuit cards with contacts, Part 3: Electronic signals and transmission protocols) and 14443-5 (Identification cards, contactless integrated circuit(s) cards, proximity integrated circuit(s) cards, Part 5: Compatibility guidelines).

In Information blocks, the Application Data field (sometimes shown as the Information/APDU field) contains application-specific information, which, thanks to the magic of the Data Link layer, is transmitted with absolute transparency.

The final field is a 2-byte CRC, used for error detection.

This concludes our discussion of proximity-coupling smart cards. We now turn our attention to the so-called *vicinity-coupling smart cards.*

VICINITY-COUPLING SMART CARDS

You will recall that proximity-coupling smart cards are designed to operate at a read distance of 4 to 10 inches from the reader. Vicinity-coupling smart cards, on the other hand, are designed to operate at a greater distance, up to a read range of 3 or 4 feet. Governed by ISO 15693, these devices are designed for such applications as controlled access cards (security cards or ID cards in office buildings or data centers), race participant identifiers (runners in marathons wear a tag on their shoe to identify the precise moment when they cross the finish line), or access control for large-scale public events like concerts, fairs, and amusement parks.

Like the standard we examined earlier, 15693 comprises four sections. Part One stipulates the physical characteristics of the device; Part Two discusses RF power, signal interface characteristics, and frame structures; Part Three, protocol details; and Part Four, the procedures used for registration of applications and issuers of devices.

ISO 15693 Part One: Physical Device Characteristics

Defined in ISO standard 7810, vicinity-coupling smart cards are identical to other card form factors, measuring 3.37 inches by 2.12 inches by 0.03 inches with specific tolerance variances. As with the close-coupling devices described earlier, the standard also stipulates tolerances for bending, heat, and exposure to radiation, UV, and X-Ray energy. In this regard the devices are, for all intents and purposes, identical.

ISO 15693 PART TWO: RF POWER, DATA TRANSFER, AND FRAME STRUCTURES

Because vicinity-coupling smart cards are designed to operate at greater distances from the reader, the air interface characteristics are different from those of the close-coupled devices. These passive devices are powered by the magnetic field of the reader, which operates at 13.56 MHz. To overcome the effects of greater distance, vicinity-coupling cards incorporate a larger antenna.

DATA TRANSFER: DOWNSTREAM, READER TO CARD

The modulation schemes used in vicinity-coupling smart cards are similar to those we studied earlier. ASK is the chosen modulation scheme, and both 10 percent and 100 percent ASK schemes are used. On top of the ASK modulation methodology is a data encoding technique known as either one-of-four code or one-of-256 code. One-of-256 code is known as a *Pulse Position Modulation (PPM) code*. PPM, sometimes referred to as *Pulse-Phase Modulation*, is a modulation scheme that uses electrical pulses of uniform height and width, but with a twist; they are displaced in time from a recognized reference position according to the amplitude of the signal at the moment the signal is sampled. PPM has an advantage over *Pulse Amplitude Modulation* (PAM); it is significantly more immune to noise than the commonly used PAM because the technique requires only that the receiver detect a pulse at a particular point in time. The duration and amplitude of the pulse are immaterial.

DATA TRANSFER: UPSTREAM, CARD TO READER

For data transfer from the card to the reader (the upstream direction), a technique known as *load modulation* is utilized. Load modulation, illustrated in Figure 2-49, is a technique in which the transmitted signal (remember, transponder tag to reader) is briefly moderated by a relatively low resistance, in this case by the transistor shown in the diagram. When the resis-

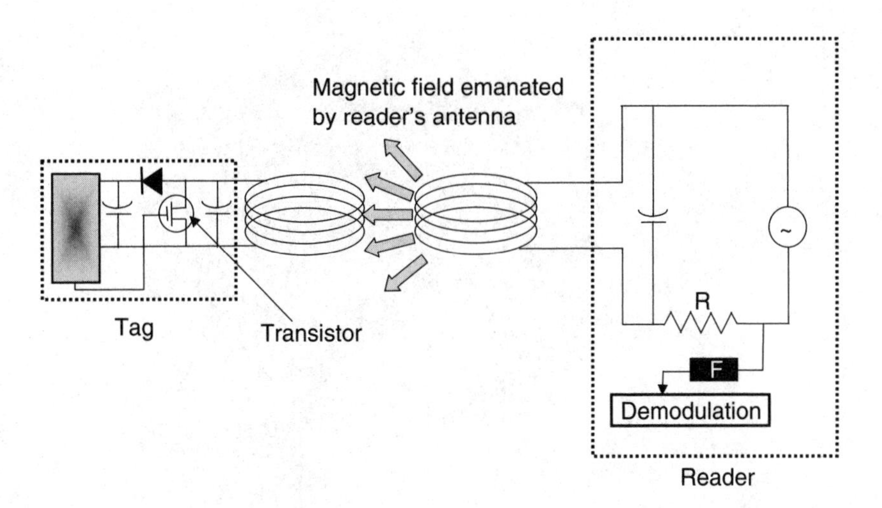

FIGURE 2-49 Load modulation.

tance increases, the current increases, resulting in a state change in the circuit. If this process is repeated at a very high rate, *subcarriers* are created that surround the fundamental signal (the carrier frequency) and that carry the desired information—the data. The trick in this well-known phenomenon is to pass the resulting signal through a bandpass filter (shown as "F" in Figure 2-49) that effectively removes everything but the desired data.

A Brief Aside: RFID Readers

Equally important as the transponder to the successful deployment of an RFID system is the reader, which energizes the passive tag and receives the information from it on demand. In some cases, the reader may manipulate the information contained in memory in the tag.

The reader, an example of which is shown in Figure 2-50, really has two responsibilities. First and foremost, it is used to activate the tags that are within its theater of control, causing them to transmit their information to the reader. Second, it serves as the interface between the theater of operations (where

FIGURE 2-50 An RFID reader. *(Photo courtesy Texas Instruments.)*

the tags live) and the system that collects, analyzes, and distributes the massive volumes of data generated by energized tags in a typical supply chain environment. The reader is often connected to a processor of some kind that collects the data before passing it on to the databases and applications that live behind it. There, the applications analyze the data, process it, and generate the stuff of management reports and dashboard displays. These applications, which have names like data mining, knowledge management, CRM, and ERP, are central to the success of any corporation; they are worth a few pages of discussion. We will discuss them here and return to our discussion of readers later in the book.

Enterprise Resource Planning (ERP)

The need to collect, analyze, and respond to knowledge about customers is of paramount importance today, because if executed properly, knowledge management can become the most critical competitive advantage a company has. This process of collecting the data, storing it so that it can be analyzed, and making decisions based on the knowledge it provides falls under a general family of processes called *ERP.* RFID systems represent a relatively new methodology for collecting this data in a mechanized fashion and passing it on to the software analysis systems that facilitate true knowledge management—and improved customer service throughout the supply chain.

ERP is best defined as a corporate planning and internal communications system that widely affects corporate resources. ERP systems are designed to address planning, transaction processing, accounting, finance, logistics, inventory management, sales order fulfillment, human resources operations and tracking, payroll, and customer relationship management. It serves as the umbrella function for a variety of closely integrated functions that are characteristic of the knowledge-based corporation.

ERP's functional ancestor was an application called *Materials Requirement Planning (MRP)*, developed in the 1970s to assist manufacturing companies with the difficult task of managing their production processes and natural resources procurement. MRP systems generated production schedules based on currently available raw materials, and then alerted management when raw materials needed to be restocked. Many barcode systems fed directly into MRP systems as containers of resources required in the manufacturing process needed to be tracked.

Over time, MRP evolved into *Manufacturing Resource Planning,* or *MRP II*. In addition to the tracking and alerting functions of MRP, MRP II allowed production managers to create "what-if" scenarios that helped them plan for unprecedented events such as a shortage of raw materials.

ERP takes the evolution to the next level by integrating financial and human resources concerns into the solution. Unlike MRP and MRP II, which originated in the production industry, ERP focuses more on the business side of the enterprise, concerning itself with the internal operational activities of the company. Clearly, RFID systems and the information they collect feed directly into this process and have the potential to offer an enormous advantage—provided there is a mechanism in place to use the information as it is collected and analyzed.

THE ERP PROCESS

During a typical business day, customer interactions yield volumes of data that are then stored in corporate databases. The data might include records of sales, product returns, service order processing, repair data, notes from meetings with customers, competitive intelligence, supplier information, in addition to the movement of products throughout the supply chain —including activities that take place at the manufacturer, the shipper, the warehouse facility, the delivery company, the advertising firm, and the market researcher.

ERP defines an umbrella under which are found a collection of functions including *data warehousing, data mining (DM), knowledge management (KM),* and *customer relationship management (CRM).* ERP defines the process by which corporate data is machined into a finely honed competitive advantage.

That data is typically archived in one or more databases in a data center. At this point in the process, the data is exactly that —an unstructured collection of business records, stored in a format easily digestible by a computer, that do not yet have a great deal of strategic value. To gain value, the data must somehow be manipulated into information. Information is defined as a collection of facts or data points that have value to a user.

Information is typically accumulated using a technique called data warehousing. Data warehouses store data, but have very little to do with the process of converting it to information. Data warehouses make the data more available to the applications that will manipulate it, but do not take part in the process of converting the data into business intelligence. In fact, corporate information systems and the users who rely on them form an information supply chain that includes the collection mechanism (among others, the RFID process), the raw material (data), the distribution systems themselves (the data warehouse and corporate network), the manufacturers or producers (the *Information Technology* (IT) organization), and the end users.

DATA MINING

Data mining is the process of converting data into information. It is the technique of identifying, examining, and modeling corporate data to identify behavior patterns that can be used to gain a competitive business advantage. Data mining came about because of the recognized need to manipulate corporate databases as a way to extract meaningful information for corporate decision-makers. The process produces information that can be converted into knowledge, which is a sense of familiarity or understanding that comes from experience or study. Data yields information; information yields knowledge; and knowledge yields competitive advantage.

Data mining relies on sophisticated analytical techniques such as artificially intelligent filters, neural networks, decision trees, and analysis tools to build a model of the business environment based on data collected from multiple sources. It yields patterns that can be used to identify new or not-yet-emerged business opportunities. Consider, for example, our earlier example of the shopper who received the miraculous phone call, inviting him to a special sale. Because of the sophisticated data collection and analysis techniques used by the retailer, the chances of a sale from that particular customer were dramatically enhanced.

Corporations that implement a data mining application do so for many of the same reasons, including the following:

Customer and market activity analysis: By analyzing the buying and selling patterns of a large population of customers, corporations can identify such factors as what they are buying (or returning), why they are buying (or returning), whether or not their purchases are linked to other purchases or market activities, and a host of other factors. One major retailer claims that its data mining techniques are so indicative of customer behavior that, given certain conditions in the marketplace, it has the ability to predict and can dramatically change the sales of tennis racquets by lowering the price of tennis balls by as little as a penny.

Customer retention and acquisition: By tracking what customers are doing and identifying the reasons behind their activities, companies can identify the factors that cause customers to stay or go.

Product cross-selling and upgrading: Many products are complementary, and customers often appreciate being told about upgrades to the product they currently have or are considering buying. Amazon.com is well known for providing this service to their customers. When a patron has made a book purchase online, he or she often receives an e-mail several days later from Amazon, telling him or her that the company has detected an interesting trend: People who bought the book that the customer recently purchased also bought the following books (list included) in greater than coincidental numbers. Amazon then makes it inordinately easy for the customer to "click here" to purchase the other book or books. This is a very good example of the benefit of data mining. This so-called shopping cart model has become the standard for online purchases.

Theft and fraud detection: Telecommunications providers and credit card companies may use data mining as a way to detect fraud and assess the effectiveness of advertising. Clearly, the data collected by RFID systems lends itself to these functions.

There is a common theme that runs through all of these subtasks. Data mining identifies customer activity patterns that help companies understand the forces that motivate those customers. The same data helps the company understand the vagaries of their customers' behavior to anticipate resource demand, increase new customer acquisition, and reduce the customer attrition rate. These techniques are becoming more and more mainstream. Furthermore, a number of factors have emerged that make the mainstreaming of data mining possible. These factors include improved access to corporate data through extensive networking and more powerful processors, as well as statistical analysis tools that are accessible by nonstatisticians because of *graphical user interfaces* (GUI), the growth of XML usage for data encoding, and other enhanced features.

A significant number of companies have arisen in the data mining space, creating software applications that facilitate the collection and analysis of corporate data. One challenge that has arisen, however, is the fact that many data mining applications suffer from integration difficulties, don't scale well in large (or growing) systems, or rely on a limited number of statistical analysis techniques. Consequently, users spend far too much time manipulating the data and far too little time analyzing it. This is one of the greatest risks with RFID systems: If the technology is deployed correctly, it will collect massive volumes of extremely valuable data. That data, however, becomes valuable only when it is converted to information and then to knowledge. A corporation that has terabytes of data in massive *storage area networks* (SANs) has nothing until it goes through the cost and complexity involved in analyzing it.

Another factor that often complicates data mining activities is the corporate perception of what data mining does. Far too often, decision-makers come to believe that data mining alone will yield the knowledge required to make informed corporate decisions, a perception that is incorrect. Data mining does not make decisions; people make decisions based on the *results* of data mining. It is the knowledge and experience of the analysts using data mining techniques that convert its output into usable information.

KNOWLEDGE MANAGEMENT

Data mining yields information that is manipulated through various managerial processes to create knowledge. Data mining is an important process, but the winners in the customer service game are those capable of collecting, storing, and manipulating knowledge to create indicators for action. Ideally, those indicators are shared within the company among all personnel who have the ability to use them to bring about change.

The collection of business intelligence is a significant component of knowledge management and yields a variety of benefits including improvements in personnel and operational efficiencies, enhanced, flexible decision-making capability, more effective response to market movement, and delivery of innovative products and services. The process also adds certain less tangible advantages including greater familiarity with the customer's business processes, as well as a better understanding of the service provider on the part of the customer. By combining corporate goals with a solid technology base, business processes can be made significantly more efficient.

KNOWLEDGE MANAGEMENT OBSTACLES

Knowledge, although difficult to quantify and even more difficult to manage, is a strategic corporate asset. However, because it largely exists in the heads of the people who create it, an infrastructure must be designed within which it can be stored and maintained, and archived in a way that makes it possible to deliver it to the right people at the right time, always in the most effective format for each user. The challenge is that knowledge exists in enormous volumes and grows according to Metcalfe's Law. Bob Metcalfe, cofounder of 3Com Corporation and coinventor of Ethernet, postulated that the value of information contained in a network of servers increases as a function of the square of the number of servers attached to the network. In other words, if a network grows from one server to eight, the value of the information contained within those networked servers increases 64 times over, not eight. Clearly, the knowledge contained in the heads of a corporation's employees multiplies in the same fashion, increasing its value exponentially. There are, however, practical limitations to this observation. Studies indicate that managers glean two-thirds of the information they require from meetings with other personnel; only one-third comes from documents and other nonhuman sources. Clearly there is a need for a knowledge management technique that employees can use to store what they know so that the knowledge can be made available to everyone.

What must corporations do, then, to ensure that they put into place an adequate knowledge management infrastructure? In theory the process is simple, but the execution is somewhat more complex. First, they must clearly define the guidelines that will drive the implementation of the knowledge management infrastructure within the enterprise. Second, they must have top-down support for the effort, with adequate explanation of the reasons behind the addition of the capability. Third, although perhaps most important, they must create within the enterprise a culture that places value on knowledge and recognizes that networked knowledge is far more valuable than standalone knowledge within one person's head. Too many corporations have crafted philosophies based upon the belief that knowledge is power, which causes individuals to hoard what they know in the mistaken belief that to do so creates a position of power for themselves. The single greatest barrier to the successful implementation of a knowledge management infrastructure is failure of the organization to recognize the value of the new process and accept it. And because the implementation process can appear quite daunting, its value is often masked by the perception of overwhelming complexity.

Fourth, the right technology must be chosen to ensure that the system works efficiently. Corporations often underestimate the degree to which knowledge management systems can consume computer and network resources. The result is an overtaxed network that delivers marginal service, at best. Today, a significant number of ERP and knowledge management initiatives fail because corporate IT professionals fail to take into consideration the impact that traffic from these software systems will have on the overall corporate network. The traffic generated from the analysis of customer interactions can be enormous, and if the internal systems are incapable of keeping up, the result will be a failed effort and degradation of other applications as well.

As corporations get larger, it becomes more and more difficult to manage the processes with which they manage their day-to-day operations. These processes include buying, invoicing, inventory control and management, and any number of other functions. Several corporations offer products that are specifically targeted at the relationships between telecommunications providers and customers. Many, for example, offer software and hardware combinations that allow customer network managers to monitor a service provider's frame relay circuits to determine whether the provider is meeting service level agreements. Given the growing interest in *service level agreements* (SLAs) in today's telecommunications marketplace, this application and others like it will become extremely important as SLAs become significant competitive advantages for service providers.

SUPPLY CHAIN ISSUES

Many of the concerns addressed by the overall ERP process are designed to deal with supply chain issues. Although supply chains are often described as the relationship between a raw materials provider, a manufacturer, a wholesaler, a retailer, and a customer, it can actually be viewed more generally than that. The supply chain process begins at the moment a customer places an order, often online. That simple event may kick off a manufacturing order, reserve the necessary raw materials and manufacturing capacity, and create expected delivery reports.

A number of companies focus their efforts on the overall supply chain. Most offer an array of capabilities including customer service and support, relationship building modules, system personalization, brand building techniques, account management, financial forecasting, product portfolio planning, development scheduling, and product transition planning.

PUTTING IT ALL TOGETHER

If we consider the entire ERP process, a logical flow emerges, which begins with normal business interactions with the customer. As the customer makes purchases, queries the company for information, buys additional add-on features, and requests repair or maintenance services, database entries accumulate that result in vast stores of uncorrelated data. The data is housed in data warehouses, often accessible through SANs.

From the data warehouses, data mining applications retrieve the data and massage it to identify trends that help the company understand the drivers behind its customers' behaviors, so that they can be anticipated and acted upon before the fact. The data mining process, then, yields information that now has enhanced value.

The information that derives from data mining is manipulated in a variety of ways and combined with other information to create knowledge. Knowledge is created whenever information is mixed with human experience to further enhance its value. The knowledge can then be managed to yield an even more accurate understanding of the customer.

Once customers' behaviors are understood, strategies can be developed that will help the service provider anticipate what each customer will require in the future. This leads to an enhanced relationship between the company and its customers, because the company is no longer in the position of simply responding to customer requests, but in fact can predict what the customer will require. This represents CRM at its finest.

SUMMARY

Let's summarize, then, the basics of these devices that are operationally based on the ISO standard 15693.

First of all, the tags are powered through reader-based inductive coupling over an air interface that operates at 13.56 MHz. From the card to the reader, signaling is accomplished using load modulation that generates subcarriers, as shown in Figure 2-51. Data is encoded using Manchester; the subcarrier is modulated with either ASK at 423 KHz or FSK at 423/485

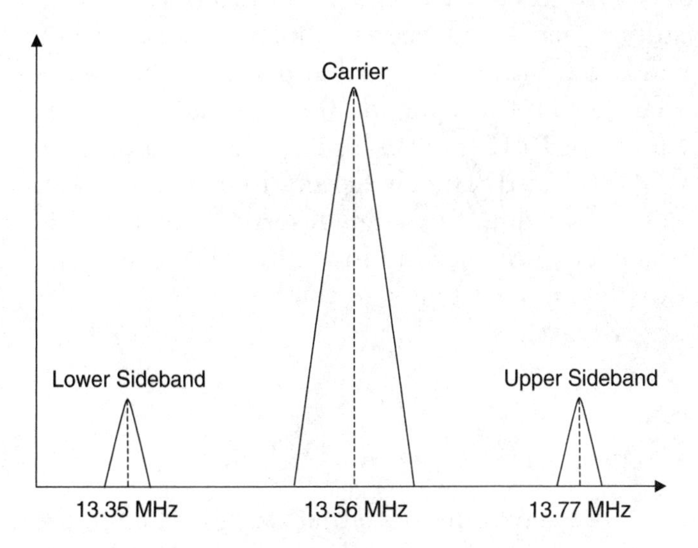

FIGURE 2-51 Subcarriers, generated on either side of the carrier signal.

KHz. The data transfer rate, which is selected by the reader based on feedback from the tag, is 26.48 kbps in fast mode and 6.62 kbps in long-distance mode. The difference between these two has to do with the modulation scheme. Ten percent ASK, combined with one-of-256 coding, works well for long-distance mode, while 100-percent ASK combined with the alternative one-in-four mode works well for short-range situations.

From the reader to the card, data is transferred using either 10-percent or 100-percent ASK, with either one of four or one of 256, as described previously. In long-distance mode, the data transfer rate is 1.65 kbps; in fast mode, the rate is 26.48 kbps.

So what have we learned about RFID systems? First, they are based on simple and preexisting technologies that have been in widespread use for quite some time. ASK, FSK, Manchester, NRZ, and radio transmission are well-understood capabilities that are entrenched in a wide range of applications. What makes them unique in the case of RFID is not what they do, but how they are being used in innovative combinations to do things in a new way. Think about it: If a warehouse manager can dramatically speed up the package handling process and improve accuracy at the same time, for less money, that's a powerful motivation to use new technology. When the hassle of torn, smudged, or lost tags goes away, when the ability to know the genetic history of a sick cow becomes instantaneous, when port security can know with accuracy that a specific container on a particular cargo ship was opened 20 hours before arriving at port, the promise of RFID and its underlying technological components will be realized. The challenges to its deployment, however, are somewhat daunting and deserve consideration. The first of these is cost; the second is security. We will examine the cost of deploying RFID later in the book; for now, let's take a look at security issues.

RFID SECURITY CONSIDERATIONS

When Dave Whitmore leaves the store that we described in the introduction after deciding *not* to buy the game console he has been lusting after, the RFID chips affixed to his pants continue to work, broadcasting his movements and whereabouts long after he leaves the electronics store. The stealth helicopters orbiting high overhead and the scruffy unbathed guy in the carpet cleaning van parked unobtrusively on the curb record everything as he moves from place to place, lodging the data in vast information warehouses for later scrutiny. They know that he buys regular gas for his car and always tops off the tank, that he buys frozen green beans instead of fresh ones, and that he likes pizza places. They know that he likes to snowboard at Smuggler's Notch, that he drinks microbrew beer, and that he and his wife like to meet friends and hang out downtown. They know how long it takes him to get home from work, how often he goes to the video rental store, how often he runs yellow lights, and what kind of dog food he buys. And worst of all? They know that he is taking—allergy medicine.

Are you worried yet about this incredible invasion of privacy that will accompany the introduction of RFID? Don't be. There are lots of things in this world to be worried about—this isn't one of them, for any number of reasons. First of all, the scenario I described in the last somewhat ridiculous paragraph doesn't require RFID: The same information can be (and is) collected through credit card purchases, ID card swipes, and execution of access controls. So if this kind of thing scares you, you should already be scared. Now I'm not trying to foment a revival of conspiracy theories, but I do want to introduce reality into the RFID equation.

Key Security Considerations

The principal purposes of data security are two-fold: to safeguard the privacy of the data that is being transmitted between communicating devices and ensure that the information travels from the transmitter to the receiver unchanged; and to verify that the sender really is who they say they are. The first of these responsibilities is handled through encryption and authentication procedures. The second is through nonrepudiation. Encryption is a technique that effectively scrambles the content of a transmitted message in such a way that it can be decoded only by someone with the appropriate key (the famous "Set Secret Decoder Ring to F" from childhood). Nonrepudiation is a technique that makes it possible for a receiver to be satisfied that the message claiming to have been transmitted by a specific source really did come from that source.

These processes are accomplished in RFID applications through the use of *Symmetrical Secret Key Cryptography,* in this case generically known as *Public Key Cryptography* (PKC). PKC is a technique that has been around for quite some time and is worthy of an explanation because of its growing role in data environments that extend well beyond RFID.

In 1976, Stanford University professor Martin Hellman and graduate student Whitfield Diffie described a concept for a secure information exchange technique based on keys, for example, a digital token that would allow a person holding the appropriate key to open an encrypted document. Shortly thereafter, MIT mathematicians Ron Rivest, Adi Shamir, and Leonard Adleman demonstrated a working model of the technique. Generally speaking the technique is known as PKC; it is also known as Diffie-Hellman, and the most common implementation of the technique is called RSA (for Rivest, Shamir and Aleman).

PKC relies on a pair of mathematically linked keys. One key encrypts the data; the other decrypts it. Interestingly, it doesn't matter which key is used first; and importantly, even though the two keys are mathematically joined at the hip, one key cannot be used to derive the other.

In RSA and other PKC techniques, the Diffie-Hellman algorithm is used to select a pair of keys. The user then advertises one of the keys, known as a *Public Key*, making it available to anyone who wants it. The other key, called a *Private Key*, remains hidden. If person 2 wants to send a message to person 1, person 2 can encrypt the information using the first person's public key; only person 1 will be able to decrypt the message using the private key. On the other hand, if person 1 wants to send a message to person 2 with a guarantee of nonrepudiation —the ability to verify that the message really did come from person 1—person 1 can sign the transmitted message using the private key. When person 2 decrypts the digital signature, person 2 uses person 1's public key, and because the message opens properly, person 2 can be absolutely secure in the knowledge that the message really did come from person 1.

Consider the diagram shown in Figure 2-52. When authentication is required (and it is not universal, by the way), the reader generates a message to the smart tag called a GET_CHALLENGE command. The tag's circuitry then kicks into action, generating a random number (Ra), which it transmits to the reader. The reader responds with its own randomly generated number (Rb). Using a predetermined secret key and decryption algorithm, the reader generates an encrypted block of data (a token) made up of a collection of control information in combination with Ra and Rb, which is transmits to the tag. The tag decrypts the token and compares the random numbers to ensure that they are the same. This guarantees that the two devices are using the same key. The tag then turns the tables on

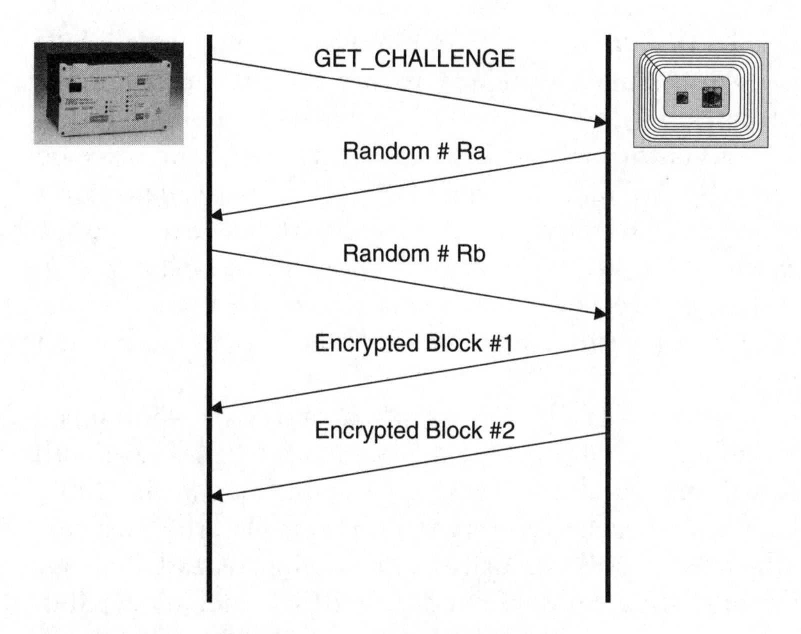

FIGURE 2-52 The encrypted token generation process used in RFID data exchanges.

the reader, generating a block of data made up of the random numbers and overhead/control information. The block is transmitted to the reader, which verifies the authenticity of the received information and therefore the validity of the session into which it is entering. Once this verification process has taken place, communications between the tag and the reader can continue securely.

Privacy Concerns

In addition to the technical concerns of adequate encryption and nonrepudiation algorithms, there is a deeper concern brewing about unauthorized scrutiny and invasions of privacy made possible by the widespread deployment of RFID transponders. Indeed, given the fact that the 915 MHz tags can be read from a distance of 10 feet or more, there is serious public concern over the ability of an electronic intruder to activate the tag on an unsuspecting citizen and download the information contained in it—or worse, track the person's movements throughout the day. So great are public concerns over privacy protection related to RFID deployment that a number of state senate bills have been introduced designed to protect the unsuspecting public from unwanted electronic surveillance. Senator Debra Bower, Chair of the California State Senate's Subcommittee on New Technologies, recently argued that RFID-marked products should not be allowed to be marketed without strong public notification of the presence of the technology, as well as a suggestion that the tags contain a mechanism for permanently disabling the transmission capability of the device once it has left the store.

The furor over RFID started when Wal-Mart announced that it planned to enter a trial deployment of RFID tags with Gillette on individual packages of razor blades late in 2003. They later scaled back their rather ambitious plan (for cost reasons—the tags are still expensive), announcing instead that they would require their suppliers to deploy RFID tracking capability at the case and pallet level by 2006. Unfortunately, Pandora's box was now open, and there was no shutting it. Adding fuel to the flames were Wal-Mart and Proctor & Gamble, which announced last summer that they had conducted an RFID trial on lipstick sold at the Wal-Mart store in Arrow, Oklahoma. Of course, the fact they also announced that secure Webcams mounted unobtrusively on the shelves allowed store managers to observe customers as they shopped for lipstick didn't help to assuage concerns about Big Brother. As a result, a large-scale trial planned by Benetton Corporation to affix RFID tags to clothing resulted in a massive protest that eventually resulted in the firm's decision to back away from RFID deployment.

Similarly, the large German retailer Metro decided in early March 2004 to stop putting RFID chips into its loyalty cards, following a storm of protest over the perceived ability of the company to spy on its customers because of the presence of the chips. These responses, and others like them, have resulted in the formation of advocacy groups like *Consumers Against Privacy Invasion and Numbering* (CASPIAN), which takes a stand against the potential for unauthorized data collection about consumer behaviors.

And of course, technology advocates have jumped into the fray with equal fervor, ponying up solutions to the perceived privacy invasion problem. RSA Security announced a blocker tag which, when placed over a standard RFID tag, effectively blocks the tag's ability to transmit its data successfully by bathing the reader in a flood of information similar to the denial-of-service attacks that hackers direct at routers. The proposed blocker tags can operate at all of the standard RFID frequencies, including 13.56 MHz and 900 MHz.

Further accelerating the move to control RFID privacy is the drug industry, which has serious concerns about customer protection due to *Health Insurance Portability and Acountability Act* (HIPAA) requirements as well as drug counterfeiting, both of which can be mitigated through the judicious use of RFID. Wal-Mart, clearly one of the most active proponents of RFID deployment, recently announced that it would require all drug manufacturers that supply prescription painkillers and other drugs subject to abuse and theft to mark the drug containers with RFID chips by April 2004. The Healthcare Distribution Management Association, which represents medical supply and pharmaceutical suppliers, has put its weight behind the technology, loudly proclaiming that the technology, in addition to protecting consumers and the industry against privacy invasion, theft, and counterfeiting, will also provide support for supply chain management and inventory control to the pharmaceutical manufacturing industry.

Next on the agenda, according to the Food and Drug Administration, are three steps: first, to purchase and deploy the various components of RFID technology and secure access to the appropriate product and customer databases; second, to move the technology into the various segments of the manufacturing and supply chains; and finally, to put into place a security philosophy that ensures the technology can adapt to changing counterfeiting measures as they occur.

There is no question that RFID is a technology whose time has come for certain applications, applications whose numbers will grow. Those applications are the subject of the chapter that follows.

PART THREE

RFID APPLICATIONS

To understand the breadth of the potential applications for RFID technology, it is important to remember exactly what RFID is—and what it is not. RFID is designed to do precisely what its expanded acronym says: *Radio Frequency Identification.* It is a technology designed to identify *things* as they pass within range of a radio-based reader, in much the same way as a barcode serves to identify a *thing* that passes beneath a laser scanner. RFID, however, has a number of advantages over barcodes. First, for the most part it doesn't matter where the RFID transponder is located on the thing that is being tracked as long as it is within radio range of the reader, which implies that package orientation is unimportant. This further implies the ability to reduce expenses related to people who must populate the supply line to orient items properly as they pass through.

Second, RFID tags are resistant to dirt, heat, paint, solvents, stomach acid, and other RFID tags, giving them a significant advantage over paper barcode labels that can be torn, dirtied, or removed in transit, making it difficult to identify a particular package.

Third, RFID tags not only deliver information on demand, as do barcodes; they also have the ability to *collect* information and store it for later review. Finally, RFID has the advantage of distance. Active transponder tags can be read from a significant distance, making them far more suitable to certain applications.

Of course, a downside to RFID exists. First, the tags can still be somewhat expensive—as much as $5 each for some types, even more for larger active varieties. No retailer is going to attach a $5 tag, or even a $1 tag, to a $3 package of merchandise. The economics don't work. Second, there remain security and privacy concerns over the perceived pervasiveness of the technology. Whether the concern about potential abuse is real or imagined doesn't matter, because a negative perception exists in the marketplace about the technology, and that perception quickly becomes reality and therefore must be managed. RFID is a powerful and capable addition to the technology collection, and offers the potential for significantly improved relationships between suppliers and customers at all levels. But like all newly introduced products, it has and will have detractors until it becomes so embedded in the public's psyche as to become invisible. Accelerating this phenomenon is the diversity and acceptance of a wide-ranging collection of applications that have a positive impact on the enterprise and residence consumer, an impact that offsets the negative feelings about RFID's perceived invasive nature. However, like all newly introduced technologies, RFID will take some time to overcome the perception that the dangers, hazards, and threats it presents are more than offset by the tremendous benefits it delivers to the marketplace.

RFID applications fall into two principal categories. The first includes short-range applications, which, as the name implies, are characterized by the need for the transponder and the reader to be in close proximity to one another, as in access-control or secure-ID applications. The second major group of applications is the medium-to-long-distance application set, which allows the distance between the two to be significantly greater, as in tollbooth and some inventory-control applications. We begin with an examination of short-range applications.

SHORT-RANGE RFID APPLICATIONS

Short-range applications imply that the transponder must be very nearly adjacent to the reader, typically with no more than a foot of read distance separating the two. These applications include those that require a user to pass a bracelet or other transponder device close to a reader, insert a card into a card reader, or otherwise be physically adjacent to the data collection device. Applications that fall into this category include the following.

ACCESS CONTROL

Access control typically involves the use of a credit-card-like device, which is inserted in a card reader. The card contains information about the card holder and is used to control access to secure areas. And although this application is not new, the degree of control it offers is. Consider the following scenario, for example. The bearer of the card, who works in a biohazard area, leaves one containment area and tries to enter a nonsecured area without activating decontamination procedures. RFID readers could ascertain the fact that the person has not passed through decontamination and would deny access to the administrative areas of the building. These applications typically involve the use of a plastic card (like the ubiquitous credit card) on which the magnetic stripe has been replaced with an RFID transponder that has been laminated inside the card, as shown in Figure 3-1. The RFID transponder inside the card is

FIGURE 3-1 RFID components laminated inside plastic access card.

secure and cannot be modified, unlike the credit card's magnetic stripe which can be erased and rewritten. And because the system is contactless, required reader maintenance is minimal.

Another access control application is being deployed by delivery services such as Federal Express. In cases where drivers need entry to buildings where there is no one to provide access, the drivers wear wristbands that have RFID transponders in them that uniquely identify the driver and grant access (Figure 3-2). For highly secure environments, the wristband devices can be combined with retinal scanners or keypads for multiple layers of secure access.

TRANSPORTATION TICKETING

Because of the number of people who use public transportation every day for commuting to and from work, demand is high for a simple way to provide access to trains, subways, and buses for high-volume commuters that does not require riders to stand in

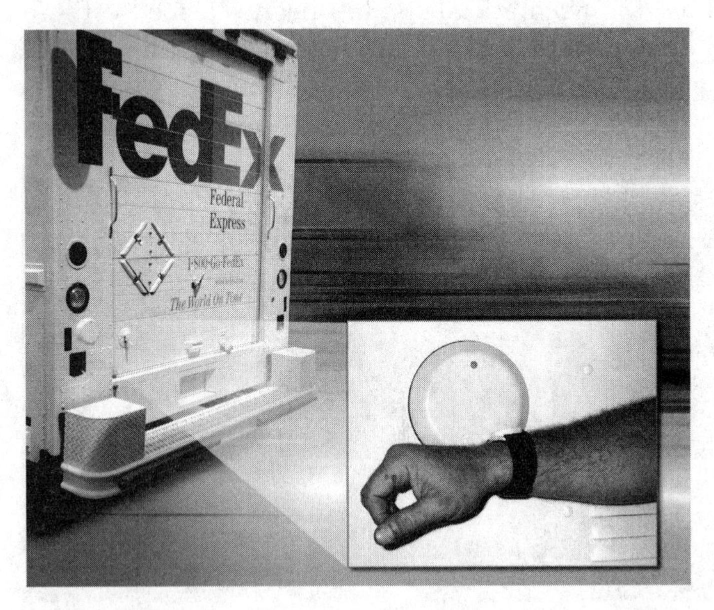

FIGURE 3-2 Wristband worn by Federal Express drivers that controls access to vehicles and secure locations. *(Photo courtesy Texas Instruments.)*

> One Saturday morning, Bob decides to help his wife with the laundry. He goes down to the basement, picks up a pile of laundry lying on the floor, and tosses it into the washing machine. He adds detergent and fabric softener and pushes the START button, which is the only button on the entire washing machine.
>
> Unfortunately, the pile of laundry that Bob tossed into the washer consists of 11 white shirts, eight pairs of white socks — and a bright red handkerchief. A catastrophe is in the making that Bob won't easily recover from.
>
> Luckily, RFID will save him. Transponders sewn into the tags on each piece of clothing identify the item's color and washing instructions and transmit the information to a reader in the washing machine. Because a conflict exists between the whites and the red handkerchief, the washer will not start. Instead, it will display a warning message suggesting that Bob find a different way to help his wife.

line to buy tickets every day. Similarly, other access applications exist to which RFID tags are ideally suited: concerts, amusement parks, state and national parks, fitness facilities, ski resorts, and other high-volume participant environments. Many are now using disposable RFID wristbands or, in the case of ski resorts, lift tickets similar to the Tyvek band shown in Figure 3-3.

FIGURE 3-3 RFID wristbands used for large event access control. *(Photo courtesy Precision Dynamics Corporation.)*

These applications of RFID have several distinct advantages. Because they are proximity devices and do not need to be physically touched by a reader, maintenance of the reader is minimal. They also eliminate the possibility of a counterfeit ticket or paper tag being used. Finally, because they are relatively low cost, they can be made to be disposable.

Personnel Identification

In the last few years, security has become a major formative force for technology deployment, including RFID. One area where the technology can be successfully deployed is personnel identification. By encoding specific information in the memory of an RFID transponder, personnel can be tracked inside a secure facility in real time. The technology can also be used to ensure that security guards are walking the entire area they are supposed to patrol and to make sure that all areas to be monitored are covered at all times.

The military is looking at RFID for personnel applications as well. For example, they are considering the deployment of button-sized RFID transponders that can be sewn into clothing. These tags will contain information about each warfighter including blood type, drug allergies, unit details, specific skills, and other data that could be useful to the logistics of warfare.

Blood, Tissue, and Organ Identification

Healthcare is becoming one of the bright spots for the introduction of RFID applications. Because of the need to minimize errors with regard to identification of patients, transplant organs, and blood, RFID emerges as an ideal solution for detailed identification. And because of the encryption capabilities of the information being transmitted between transponders and readers, HIPAA requirements are met. Today, RFID tags can be placed on sample collection bottles that identify the source of the sample, the time it was collected, where it has been between the patient collection point and the lab where it will be analyzed, and so on. Tissue and organs can be similarly identified for cross-matching purposes to guarantee that the tagged item is used appropriately to avoid rejection.

Precision Dynamics Corporation (PDC), working with Georgetown University Hospital's blood bank, recently began a pilot study to determine the degree to which RFID wristband solutions might increase the safety and reliability of blood transfusions. PDC manufactures the Smart Band© RFID Wristband System (Figure 3-4), around which the study will be conducted.

For approximately two years, the hospital's Outpatient Infusion Service has used PDC barcode products to verify the accuracy of blood transfusions. And although the technique has worked well, the hospital believes that RFID may offer a more efficient, effective, and safe alternative. PDC's Smart Band serves as a small dynamic database that stores an individual patient's information during his or her stay in the hospital.

PDC's Smart Band relies on an embedded Texas Instruments Tag-It© RFID solution. The peel-off, writeable RFID labels are printed and encoded using a Zebra Technologies

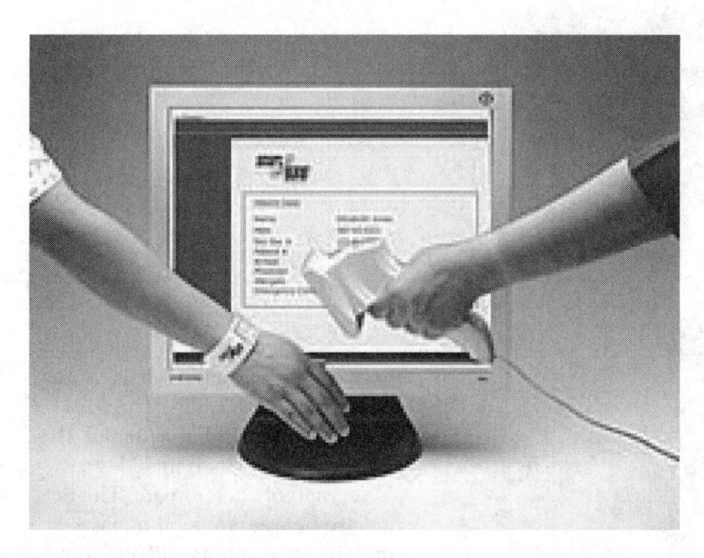

FIGURE 3-4 RFID-based patient identification band, which serves as a portable patient information database. *(Photo courtesy Precision Dynamics Corporation.)*

R402 RFID printer, and the patient database system is provided by AMTSystems PatientSafe™. The initial phase of the study began on March 1, 2004, although the research and clinical implementation phase will begin two to three weeks after completion of the initial phase. The study will evaluate and compare the effectiveness of barcode vs. RFID solutions for blood transfusions of 100 patients.

INMATE IDENTIFICATION AND ACCESS CONTROL

Another successful application for RFID wristband technology is inmate tracking in prisons (Figure 3-5). PDC's Clincher Smart Band product line provides the basis for this particular application. The wristbands are encoded with each inmate's personal data. When the inmate walks through a portal, a 13.56 MHz reader mounted on the wall reads the transponder in the wristband. The information in the tag is then indexed to a cen-

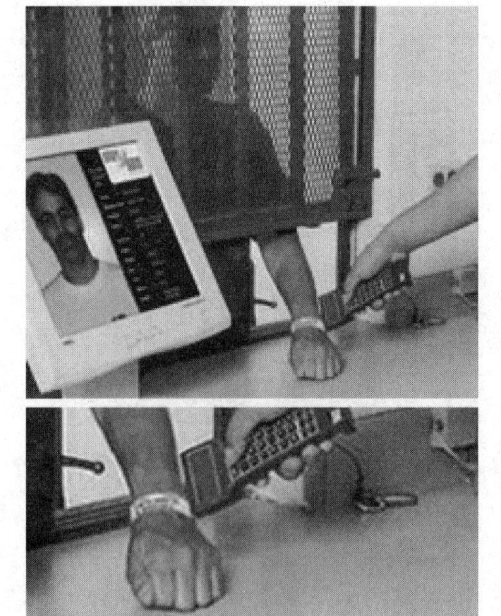

FIGURE 3-5 Inmate identification in a correctional facility. The RFID tag in the wristband must match the database entry, which includes a photograph of each inmate. The tags can also be used to debit inmate commissary accounts. (*Photo courtesy Precision Dynamics Corporation.*)

tral database to instantly retrieve and verify the individual's photo before granting access. The Smart Band can also be configured to debit each inmate's commissary account each time purchases are made.

Hazardous Waste Monitoring

Because toxic waste is so carefully controlled in terms of its storage, identification, handling, and ultimate disposal, the traditional barcodes used for marking containers are in many ways inadequate. It is far too easy for them to be torn, smudged, or otherwise rendered incapable of conveying data to a reader, and for a cargo as critical as hazardous waste, this is not acceptable. Therefore the electronic transmission of data that RFID enables is a far better way of conveying the information required to handle the cargo safely.

Fleet Management

One of the challenges of operating a rental or company car fleet is inventory management. How often have those of us who are travelers been given the space number for our rental car, only to discover after trundling our bags across the parking lot that there's no car there? RFID is being deployed by fleet operators around the world to eliminate that problem. By installing transponders in each vehicle, which can be read when they are parked in a parking spot, inventory control is dramatically simplified. Alternatively, the situation can be reversed. Instead of installing a transponder in each car and a reader in the fleet yard, some commercial fleet operators are installing the reader in the car and the transponders throughout the operating area of the vehicle. In this scenario, medium-range tags are mounted in locations throughout the delivery geography of the vehicle. As each vehicle passes within range of the tag, the time and location is recorded in the vehicle's reader database, allowing logistics managers to improve routing and delivery models.

Vehicle Identification

The application seems obvious: Use RFID technology placed secretly in vehicles to identify them in much the same way that a *vehicle identification number* (VIN) is used today. This application takes on several forms. First, it can be used for access control, as shown in Figure 3-6. As cars pass the reader mounted on the arm, their on-board transponders are scanned and entry is allowed if appropriate.

Related applications include keyless entry, which opens the locks of a car when the appropriate transponder is within proximity of the vehicle; and antitheft, which prevents the car from being started without the appropriate RFID-equipped key.

Pigeon Racing

OK, so it's not the most common of applications, but in some countries it's as widely observed as NASCAR is in the United

FIGURE 3-6 Automated vehicle identification, here used at McDonald's. (*Photo courtesy Texas Instruments, Inc.*)

States. In this sport, homing pigeons are carried far from their homes and released. Their return is then carefully timed. For this application, companies like SOKYMAT manufacture special transponders that are affixed to each bird's leg. When the bird arrives at its destination, its precise arrival time is recorded by the reader that is installed at the bird's coop. Examples of these rings are shown in Figure 3-7. SOKYMAT is purportedly the single largest manufacturer of RFID transponders in the world, producing more than 50 million devices annually.

PRODUCTION LINE MONITORING

Remember the fishbone diagram that we discussed earlier with regard to the value chain? RFID has the potential to play a large and important role in the creation of value in the production line environment. As products are incrementally assembled, tested, and packaged, they move through the production line at a certain pace. A certain percentage of the products are typically removed due to a manufacturing defect or some other factor, although the remainder move through the process unencumbered and are loaded for distribution to intermediate warehouse

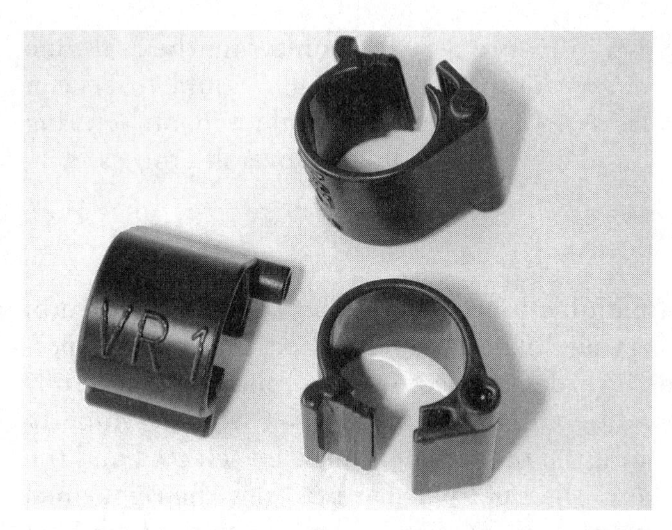

FIGURE 3-7 Pigeon racing rings. *(Photo courtesy SOKYMAT Corporation.)*

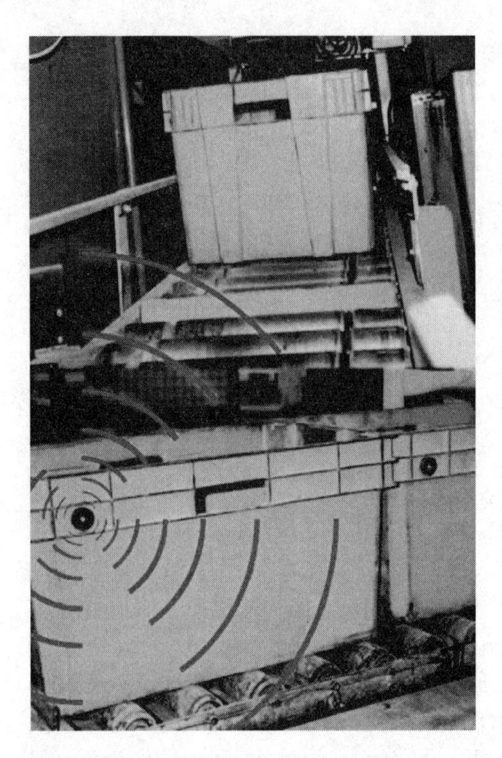

FIGURE 3-8 Tracking products along the supply line. *(Photo courtesy Texas Instruments.)*

facilities, as shown in Figure 3-8. By monitoring the real-time progress of product creation and movement, manufacturers can tighten supply chains, add efficiency to product manufacturing and distribution, and mine cost out of the overall process.

Car Body Production

When cars are manufactured, the number of steps involved in the process is dizzying. Like a complex dance, they must be performed in a specific order, at specific times, and under carefully orchestrated conditions. For example, when it comes time to paint the car's body, the proper color must be selected and the car body must be in the right place at precisely the right time. Using RFID transponders that are impervious to the high heat used in the painting process, the proper information is

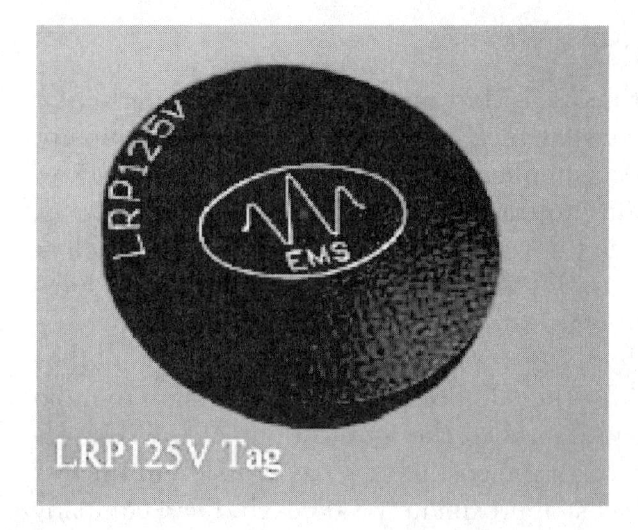

FIGURE 3-9 A high-temperature RFID tag attached to a tire prior to entering the painting oven. *(Photo courtesy Escort Memory Systems.)*

conveyed to the robotic painters along the assembly line so that the process works flawlessly. The tags, shown in Figure 3-9, are typically attached to a tire or to the underside of a wheel well.

PASSPORT SECURITY

The European Union has already made a commitment to require that passports have RFID devices embedded in their covers in compliance with the United States mandate that visitors' new passports be RFID-equipped by October 2004. The RFID tags can carry digital photos and fingerprints, entry visa information, and other data that is critical for immigration officials.

LONG-RANGE RFID APPLICATIONS

Long-range (or longer-range) applications typically rely on active, battery-powered transponders because of the need to successfully transmit data over a greater distance. The most common of these are described here.

SUPPLY CHAIN MANAGEMENT

Everyone knows that Wal-Mart set the bar for this particular application. Wal-Mart, which is the world's largest company, gave its 100 largest suppliers a deadline of January 2005 to implement the ability to track their products using RFID tags. The deadline, handed down in November 2003, made it clear that suppliers that comply will be treated preferentially—a message that is hard to ignore.

One problem that faces suppliers is that to a large extent they have not budgeted for the additional cost of conversion to radio tags, nor have they researched the implications of their implementation. Even though the cost of tags has dropped in the last few years, they still cost approximately $.05 each. For a company that supplies millions of products a year to Wal-Mart, that quickly adds up, even at a few cents each. For that reason, Wal-Mart modified its original request, asking suppliers to tag products at the pallet level rather than at the individual product level. As costs continue to drop, they may change that request, but for now pallet-level tracking is adequate for their requirements.

Another key issue is that Wal-Mart is still refining their technical requirements. They have requested that suppliers use Class One, Version Two tags, which are read-write devices. Unfortunately they are not yet widely available in the writeable form.

The bottom line is that analysts estimate that consumer goods suppliers will spend between $13 million and $23 million each in 2004 to comply with Wal-Mart's mandate. That's a significant investment, but one that Wal-Mart and its suppliers believe will pay off many times over. Tom Williams, a spokesman for Wal-Mart, told the press that RFID will make it possible for Wal-Mart to increase the availability of products in stores to very nearly 100 percent all of the time, an increase from 99.3 percent in the pre-RFID timeframe. And although this doesn't seem like a huge number, remember that we're talking about Wal-Mart. The difference between 99 and 100 percent in stock is $1 billion in sales revenue.

Nevertheless, at the time of this writing, Wal-Mart's plan that calls for its top 100 suppliers to reach a point where they can guarantee 100 percent readability of RFID tags at the case and pallet level isn't moving along at the pace they would like. According to a report issued by Forrester Research, less than 25 percent of Wal-Mart's suppliers will meet the January 1, 2005 deadline. And the problem isn't simply cost, although that's a major contributor: The average RFID startup cost for Wal-Mart compliance, including a year of technology maintenance, is approximately $9 million.

The bigger problem is technology readiness. For most companies this is uncharted territory. Few pre-existing business cases can exist to use for benchmarking their deployment plans, resulting in a slow and sometimes chaotic conversion process.

According to Forrester, companies must spend as much as $100 million to see real benefit from RFID. They believe that a technique known as *source tagging* is the best way to achieve success with RFID implementation in the supply chain. Source tagging involves placing RFID tags on finished goods at the manufacturing facility, then tracking them through the entire supply chain. Because of the cost of doing this, only the largest suppliers are in a position to do so. Many of Wal-Mart's suppliers simply won't be able to come up with the cash for such a conversion and will therefore take longer to meet the company's mandate.

Most analysts believe that Wal-Mart will be forced to back down from its demands and refine its requirements if the RFID deployment is to succeed. The technology has the potential to be exceptionally valuable in the supply chain world, but most agree that it simply isn't quite ready for a full-blown implementation on the scale that Wal-Mart is looking for. However, given the pace at which technology advances, the timeline for deployment will be compressed and the technology will be deployed fairly soon. Most likely, however, it will be deployed in stages.

In April 2004, the company started an RFID pilot with its pharmaceutical suppliers. And although all of the suppliers are not yet in compliance, enough are to engage in a valid trial.

There is no question that RFID has an uphill implementation battle to fight for a number of reasons including the cost of deployment, changes in internal system philosophies, and conversion from an entrenched, fully functional product identification technique (barcodes) to RFID tagging. In spite of these challenges, however, most companies in the industry agree that the technology will provide tremendous advantages with regard to cost reduction and product delivery and availability.

So what is the current state of the art with regard to supply chain implementation? In the grocery industry, supermarkets tag pallets, cases, and other *returnable transit containers* (RTCs), such as the plastic crates used for the transport of meat and produce. RTCs represent expensive capital assets that have an annoying tendency to disappear, so the ability to see and track them is essential for proper management of the RTC asset pool. Furthermore, the ability to write to the transponder allows information about the contents of the pallet, the sell-by date, and the manufacturer to be encoded, ensuring compliance with product delivery requirements. By linking the data contained in the tags to inventory management systems, suppliers can ensure that products are moved to the shelves in strict rotation to prevent spoilage and minimize out-of-stock situations.

MAIL AND SHIPPING

At the 2004 CeBIT show, the Swedish Cypak Corporation demonstrated a technique for using RFID in packaging. They showed the technology used in a package that contained medicinal tablets. When a tablet is popped out of the package, the package records when the pill was removed and allows the recipient to register details about his or her experience with the medication, such as how well he or she slept while using it and whether he or she suffered any side effects. When the packet is placed on a reader connected to a PC, the information captured by the smart package is automatically transmitted to a physician or pharmaceutical firm via e-mail.

This same packaging technique has ready applicability in the shipping and postal sector, illustrated in Figure 3-10.

FIGURE 3-10 RFID-marked packages being shipped through a high-speed transit center. *(Photo courtesy Texas Instruments.)*

Cypak's technology could be used to determine whether a package has been opened during transit, which could indicate that the product was stolen or tampered with. It could also indicate whether products were kept at an appropriate temperature, whether fragile items were handled appropriately, and whether the products carried within were tampered with. The natural and obvious extension of this is security tagging; RFID seals are now being created and deployed for use on large shipping containers. If the containers are opened during their transoceanic voyage, something that should not occur under any circumstances, the RFID-based seals record the event and whatever circumstances related to it that they are programmed to collect, and transmit them to a nearby shipboard reader for transmission to a data collection system for notification and analysis.

Clothing Tags

The incorporation of RFID tags into garment labels or even woven into the garment itself can be a valuable tool for clothing brand owners. A tag that is added to the garment at the manufacturing facility, for example, can identify the source of the garment. By using the tag's unique identification number, the garment can be certified to be authentic, allowing for identification of and control of the distribution of counterfeit or gray market products. The World Trade Organization (WTO) has expressed interest in any technologies that can be used for this purpose.

Many companies have begun extensive trials of RFID for garment tracking, including such names as Marks & Spencer. The London-based retailer conducted an early trial of the technology at its High Wycombe store near London with great success and has now indicated its desire to expand its use of the technology. The trial, funded partially by the Department of Trade and Industry, involved the use of RFID tags on men's suits, shirts, and ties. A portal was installed at the distribution facility and loading bay through which racks of garments and packaged clothing were pushed and read at a high speed. Similarly, a mobile scanner attached to a shopping cart was used to scan garment tags on the shop floor. And although Marks & Spencer concluded that the portal scanner was somewhat less accurate and versatile than the mobile scanner, the overall trial was successful. During the trial, 50 customers browsing the menswear department were interviewed about RFID, and they expressed no issues with (or frankly, awareness of) the use of RFID for product tracking.

To avoid the privacy issues that have cropped up in many RFID trials, Marks & Spencer used intelligent labels that only contained information specific to the garment itself—size, style, and color.

Other garment-related uses for RFID technology include antitheft devices such as those shown in Figure 3-11. One interesting and recently announced sideline technology is called *Chipless ID*. Tapemark, a manufacturer of packaging and packaging materials, is introducing it as a way to embed RFID transponders in packaging materials. Invisible to the eye, each transponder emits a unique signal that cannot be forged, making the technology particularly suitable for authentication, anticounterfeiting, and security applications.

FIGURE 3-11 A lightweight garment tag that will trigger an antitheft device if a tagged garment is taken from the store.

Unlike RFID chips that contain an antenna, a processor, and associated memory, each chipless ID transponder consists of nothing more than a passive antenna. The antennas are manufactured from very small (5 microns in diameter, 1 mm in length) fibers called *nano-resonant structures* that can be detected by special readers. The readers transmit a coherent pulse toward a transponder and receive an interference pattern in return that is processed and identified. The frequencies used range from 24 GHz to over 60 GHz, a range that is well above conventional UHF frequencies and the 2.4 GHz microwave frequency established for use by standards setter EPCglobal. And although the technology clearly can be used for supply chain applications, immediate interest in the technology is for security and tracking applications.

One advantage of this new chipless technology is cost. Tapemark believes that they will be able to produce tags that cost less than 50 percent of the cost of RFID chip-inclusive tags. According to company sources, chipless ID tags will soon be available at prices that fall below $.05.

Using chipless technology, Tapemark embeds fibers in paper labels to create passive antennas that are undetectable, making them ideal for authentication or product security applications. Because each antenna creates a unique interference pattern, they can be assigned unique serial numbers that can be tracked from product creation through distribution. The read range will most likely be no more than a few inches.

Applications for Tapemark's chipless ID solution are exciting. The technology will help to ensure that a product or package is authentic and secure by adding RFID to the product label or package. The U.S. *Food and Drug Administration* (FDA) has recommended that manufacturers employ anticounterfeiting technology to ensure they are genuine, particularly for drugs and other pharmaceuticals. The technology, which Tapemark licenses from an undisclosed third party, is ideal for this application because the spatial pattern of the antenna in each transponder is randomly generated when the tag is created and is therefore unique, which means that the signal it generates when interrogated by a reader is equally unique. Furthermore, that signal is converted to a random number by associated security software, and because the number is random, it cannot be copied. Tapemark is targeting the medical, pharmaceutical, and consumer products industries as potential markets for its technology.

Library and Rental Companies

Many libraries use RFID to automate the loan and return of books, tapes, DVDs, and CDs and to provide a mechanism for the creation of a real-time library inventory. Historically, library properties were tracked using barcodes, each of which had to be read by a reader—a highly human-intensive process. Using RFID tags such as the sticky label shown in Figure 3-12, library holdings can be checked in and out automatically and inventory processes can be automated using shelf-mounted or handheld scanners. This results in a reduction in the need for personnel and a much higher degree of accuracy in inventory management.

Baggage Handling

Because of increasing volumes of handled baggage and parcels, airlines have conducted RFID trials over the past few years to determine whether the technology lends itself to improvements in efficiency and accuracy. The tests have been successful:

FIGURE 3-12 A sticky RFID tag of the type used for inventory control in libraries.

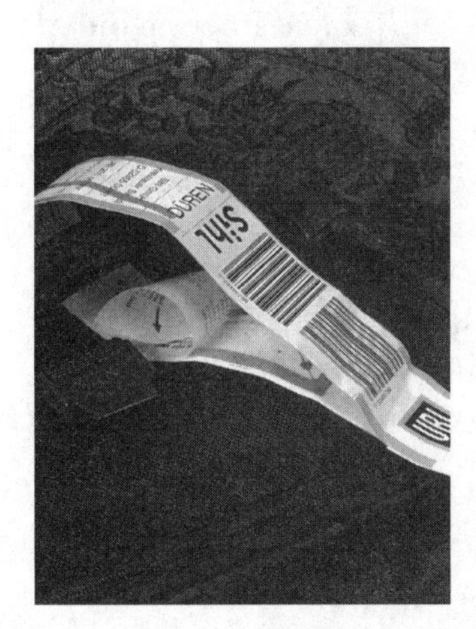

FIGURE 3-13 RFID-equipped luggage tag: Airlines using these devices show an increase from 90 to 99 percent in baggage handling efficiency. *(Photo courtesy Texas Instruments.)*

First-read rates of over 99 percent were recorded using RFID tags like the one shown in Figure 3-13, compared to read rates of less than 90 percent for barcodes.

By adding read-write capability, the information contained in the tag can be changed along the way as the bag or parcel makes its way through the handling system. The value of this capability is clear: Bags can be held for security screening and then released for loading once they have been checked, with the RFID database monitoring the information for agreement.

FOOD PRODUCTION CONTROL

Because of the enormous volumes of foodstuffs produced within the country and imported from other countries for local consumption, it has become particularly critical to monitor their movement from the point of production (import/export dock, feedlot, agricultural concern, freezer facility, and so on) to the point of consumption (grocery store, restaurant, or consumer's home). Adding to the urgency are recent concerns about preventing the spread of Bovine Spongiform Encephalopathy (Mad Cow Disease). The recent situation in the northwestern United States, where a cow was found to have the disease but its point of origin proved difficult to determine, could have been avoided entirely through the use of an RFID transponder carrying information about the cow's origins, medical history, DNA, and so on (see Figure 3-14). And although it has not yet happened, expect legislation in the near future that mandates electronic records for food in the United States and, most likely, other countries as well.

Alternate animal tagging techniques, discussed earlier, include the glass transponders shown in Figure 3-15, which are injected subcutaneously using the disposable syringe shown in Figure 3-16.

FIGURE 3-14 RFID tags in cows' ears. The tags can be used to carry medical history, origin information, and other data critical to the safety of the nation's food supply. *(Photo courtesy RFID Journal.)*

FIGURE 3-15 Glass transponders that are injected subcutaneously for livestock identification.

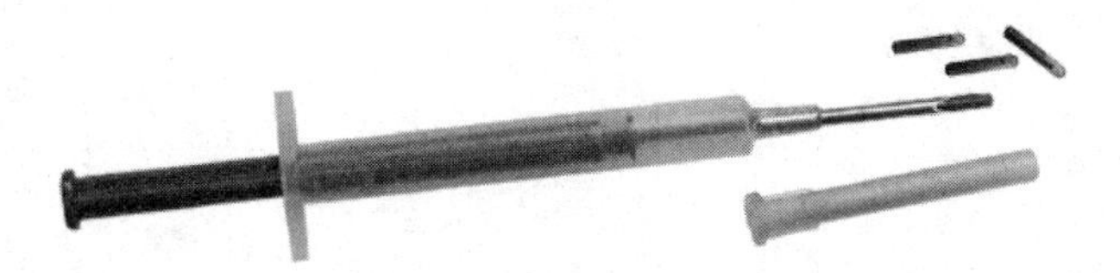

FIGURE 3-16 Disposable syringe used for subcutaneously injecting glass transponders shown in the upper right of the photograph and in Figure 3-15.

The applications described here represent a sampling of the capability that RFID and the databases that process the data it creates will enable. Although challenges will continue to present themselves related to cost, privacy, security, spectrum availability, data analysis, and knowledge management, and the challenges associated with the unseating of a fully functional incumbent technology in favor of a new one, RFID will enjoy significant successes. As we observed earlier, all new technologies that introduce change experience resistance at various stages along the way and RFID is no exception. Invariably, they overcome those challenges and demonstrate the value that they can bring to business and society, and the revenue and cost savings associated with their deployment start to flow. Many companies have jumped on the RFID juggernaut; in the next section of the book, we look at them and at the products they bring to the marketplace.

PART FOUR

FINAL THOUGHTS

In this final section, we look at RFID from a slightly different perspective. We've already considered the inner workings of the technology behind RFID and the special challenges it presents, the typical applications that have emerged to take advantage of RFID, and the history that led to its development and eventual deployment. From a technology perspective we have discussed transponder tags, readers, and the radio link that exists between them. We have discussed the applications that are positioned to take advantage of the data that will be generated in enormous volumes by a full-blown RFID implementation, and the ultimate supply chain advantages that will result.

Now, we turn our attention to the practical aspects of RFID —implementation considerations, emerging real applications, and a bright-light-of-day view of what RFID really is—and more importantly, what it isn't. We'll also take a look at the companies that are manufacturing in the RFID game and the specialized products they are creating for the market. We've already looked at Texas Instruments, one of the early aggressive supporters of RFID, but they are far from alone. Infineon, for example, has been in the game for a very long time, as has SOKYMAT. The list gets longer every day.

RFID IN ACTION

Let's begin with a discussion of the realities of RFID. It is an asset tracking technology—no more and no less. Like a two-dimensional barcode, it has the ability to track the location of an asset as the asset makes its way through the supply chain. It is not designed to track unsuspecting people, nor does it represent any more of a privacy invasion than credit cards do. In fact, there is a global system in place already (and has been for years) to enable the use of credit cards anywhere in the world. There is no such system in place for RFID. Furthermore, credit card purchases result in a record of monthly acquisitions, data that can be used to create a very accurate profile of the card's user. No such capability exists for the use of RFID in that regard. So first and foremost, let's dispense with the hype about RFID serving as the next Big Brother invader.

Let's examine a typical RFID system implementation by following a product from its point of manufacture to the point at which it is delivered. For the purposes of our discussion, let's assume that the product is a portable DVD player. Attached to the inside of the box in which the device is shipped is a passive RFID transponder embedded in a sticky label such as that shown in Figure 4-1. The transponder comprises the antenna and the microchip and memory, which contains an *electronic product code* (EPC). The EPC is readable by either a fixed

FIGURE 4-1 An RFID transponder attached to a sticky label that can be affixed to a dry surface for shipment.

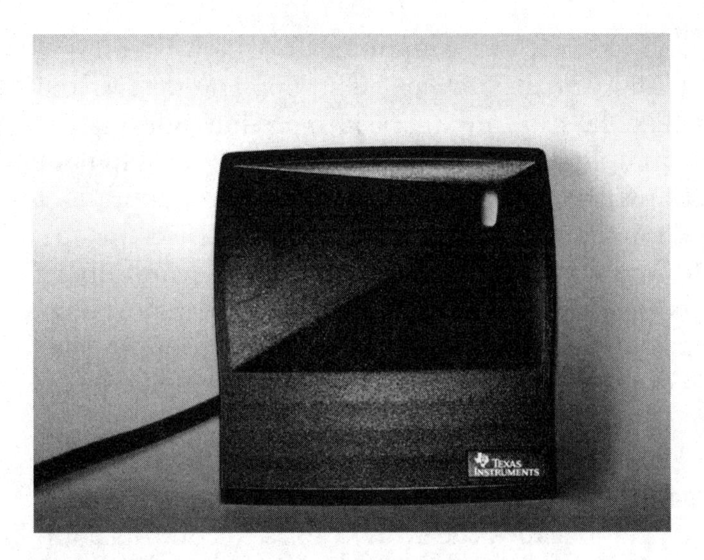

FIGURE 4-2 A fixed RFID reader. *(Photo courtesy Texas Instruments.)*

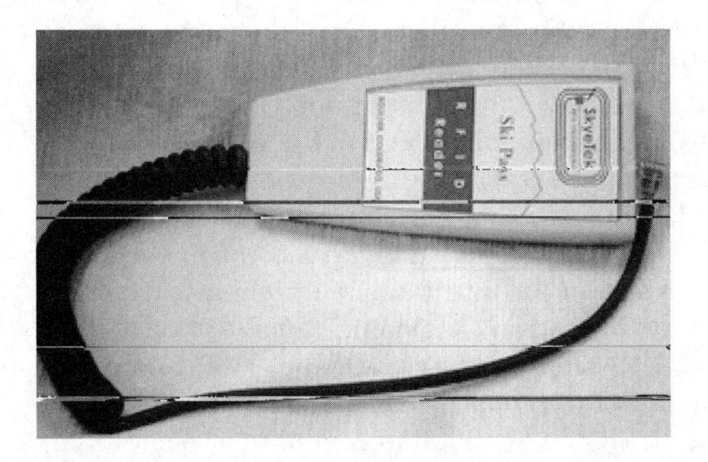

FIGURE 4-3 A handheld RFID reader. *(Photo courtesy SkyeTek Corporation.)*

reader (Figure 4-2) or a hand-held device (Figure 4-3). When the tag attached to the DVD player comes into the radio field of the reader, the tag is activated by the RF energy emitted by the reader, causing it to transmit the EPC embedded in memory. The reader receives the EPC over the radio link between the two devices and passes it to a computer, in which is running supply chain management software. The application has the ability to recognize the EPC and relate it to product information such as individual unit price, manufacturing and shipment dates, instructions for care (for example, because the product we're talking about is a DVD player, it makes sense that we would want to know that the package is treated carefully in keeping with its fragile contents), and other special instructions germane to each product and tag. The tag information, in concert with the computer and its applications, are used to track the product throughout the supply chain.

As the DVDs are placed into boxes and multiple boxes are bundled into pallets, the RFID tags are affixed so that the boxes can be tracked. And because the RFID protocol controls collisions, the fact that there are multiple tags emitting signals simultaneously does not present a problem to the reader whose field they are in. The pallets are shrinkwrapped and carried out to trucks that are waiting to deliver them to their next destination. As the forklift carries the pallets out of the manufacturer's facility, a reader at the exit portal records their departure.

The pallets are trucked to a distribution center where they are removed from the manufacturer's truck and carried into the building. As they enter, a reader at the entry portal records their arrival and updates the on-hand inventory database. No human intervention is required and no packages need be opened. The packages are stacked in the intermediate warehouse according to the order of their next trans-shipment, which will carry them to a retail store. Ultimately they are placed on an outgoing truck and shipped to the various retail establishments where they will be sold. As they leave the distribution center, their departure is once again recorded and the supply chain database is updated.

As the DVD players arrive at the retail store, their arrival is noted by a reader mounted adjacent to the back door through which they are carried into the store. The pallets of product are left in the back room where they are unpacked. Some of the DVD players go immediately out onto the floor for sale; the others are placed on shelves and constitute on-hand inventory.

On the retail floor, the new DVD players are placed on display racks where a clerk with a handheld reader records the devices that have been placed for sale.

When a customer purchases one of the players, the checkout clerk interrogates the tag with a handheld reader and generates a record of sale. The EPC information is automatically applied against the database of on-hand inventory. At the same time, the tag on the sold device is deactivated so that it will not set off alarms when the consumer passes through the antitheft portal located at the front door of the store. The data can also be used to provide the customer with enhanced warranty coverage.

Meanwhile, the backroom inventory applications are hard at work, analyzing the massive volumes of sales data from all of the stores that are actively selling the DVD players. Relational databases have the ability to generate reports that indicate sales trends, spending patterns, product association indicators (people who buy DVD players often buy DVDs at the same time, so place them physically adjacent to each other in the store to make the association easy for the customer), and so on. Production levels can be managed to facilitate the success of just-in-time inventory control, ordering processes, and pricing. From one end of the process to the other, RFID provides a discrete level of control for manufacturers, shippers, retailers, and customers.

The idealized example described in the preceding section illustrates the utility that RFID brings to the supply chain. It is unquestionably a powerful, utilitarian technology that, if deployed properly, yields serious benefits. Like all new technologies, however, it must be deployed with forethought and careful planning if its benefits are to be realized and the concerns over its potential abuses of privacy allayed through a clear demonstration of its value to the corporation. Like all technologies, RFID isn't for everything. It is not, nor was it ever, intended as a wholesale replacement for barcode technology. It is designed to serve as a rich complement to barcodes, and to augment the process of tracking assets throughout the overall supply chain. They key here is that the question of whether or not to implement RFID in an enterprise has very little to do with the technology itself. It has everything to do, however, with the business.

When *chief information officers* (CIOs), *chief technology officers* (CTOs), and other officers—the so-called C-Levels—consider an investment in a new technology, they do so with precious little concern about the technology itself in terms of how it works. Their concerns revolve around a very different set of questions: Will the investment I'm about to make in this new technology help me lower my CAPEX and OPEX? Will it somehow help me to move my revenues in the upward direction? Will it help me either preserve or improve my firm's competitive position relative to its peers? And finally, will it somehow help me to mitigate the downside risk I face in the marketplace? If the new technology provides the right answers to one or more of those questions, officers will pay attention. And because officers of the corporation are those that hold the purse strings, it is critical that proposals for technology implementations be couched in terms of those questions. RFID, if done properly, positively affects every one of them. Note that the questions address more than financial issues; they also direct themselves to overall business processes.

A series of basic questions must be asked before implementing RFID in an enterprise. First and foremost are those that determine whether RFID is even a good idea in a particular business. For example, what is the data that we intend to collect using RFID transponders? What is the nature of the environment in which the data is to be collected? Is it a chemically or temperature-hostile environment? Will the tags that are to be read be affixed to objects that are moving, such as on a transport belt within a warehouse? If they are moving, how fast are they moving? How close can the readers be placed to the objects as they make their way through the environment? What, therefore, is the required read range between the transponders and the readers? Are there requirements for unattended operation, or will there be adequate available to help the process along? How often will data need to be collected and analyzed in a particular environment? Some companies have

FIGURE 4-4 Nail-shaped transponders from SOKYMAT Corporation.

designed innovative approaches to affixing tags as a way to collect data in unusual ways. Consider the nail-shaped tags shown in Figure 4-4 from SOKYMAT Corporation. They can be hammered into a wall or shipping box for immediate and semipermanent tracking.

Once the decision to deploy RFID is made, another set of questions arises that has more to do with the physical environment in which the technology is to be deployed and less to do with the overall question of whether or not to deploy RFID. For example, does the intended application require both read and write capability, or is it enough to simply be able to read the information transmitted by the transponder tag on demand? Where will the tag be physically attached to the pallet or container? Will it be placed in a location that will make it difficult to be read? Furthermore, will the tag be affixed to products that contain metal (which blocks the signal) or liquids (which absorb the signal)? Is the environment such that the surface to which the tags will be affixed becomes dirty, greasy, or covered with paint, or in some other way makes the attachment of a transponder tag challenging? Are the products being shipped packaged in such a way that multiple tags will enter the field of influence of a reader simultaneously, requiring implementation of the anticollision protocol?

Other considerations are most certainly economic in nature. RFID is not cheap: Full implementation for a large corporation can become a multimillion dollar venture, and although it will offer an attractive payback if implemented for the right reasons, it is still a significant financial undertaking. It is therefore critical to ask questions related to cost justification. The first of these, naturally, is this: What is the cost of NOT implementing RFID? In other words, if we choose not to implement, will the firm's profitability decrease over time as a result of the decision not to make the investment in the technology? Furthermore, will the firm find itself without a critical competitive advantage if its competitors decided to go forward with the implementation and it turns out to be a powerful addition to their supply chain? Will the lack of RFID result in the inability to collect important customer data that facilitates the maintenance of a competitive marketplace advantage? And as RFID becomes more commonly deployed, will existing and potential new customers see the lack of it in the firm to be a disadvantage, leading to their dismissal of the firm as a supplier?

On the other hand, if the decision is made to go forward with the implementation, will RFID's data collection capabilities, in combination with a back room analysis application, yield an improved understanding of customer requirements, better supply chain management, stronger customer loyalty, better corporate brand recognition, and more complete customer service?

Fundamentally, the key question that needs to be asked—and answered—is this: What is the problem I'm trying to solve by making this technology investment? As we stated earlier, RFID is an asset tracking technology that facilitates improvements in such areas of the business as security, logistics management, warehousing processes, inventory control, and distribution. In combination with an array of sensors connected via a sensor network, RFID provides discrete and far-reaching capabilities within the asset tracking environment. According to

Zebra Technologies,[1] a leading supplier of thermal on-demand barcode printers and RFID smart label compliance solutions, shipping errors alone cost between $60 and $250 to resolve. A large corporation, therefore, will find shipping errors to be a costly problem. A one percent savings (far less than what a well-designed RFID solution would bring) would produce savings of between $15,600 and $65,000 annually—a not insubstantial figure.

RFID IMPLEMENTATION ISSUES

There are a number of implementation issues that fall under the general category of management matters. The first of these deals with issues associated with the wireless link between the reader and the transponder. Radio waves can be disrupted by any number of causes in the environment, resulting in unpredictable dead spots. Unless the system is carefully designed, the same can happen with RFID. Furthermore, because metal tends to block the RF signal used in RFID and liquids absorb it, unanticipated service gaps can occur in warehouse, distribution center, or manufacturing facilities, leading to improper data collection. And because certain frequencies have a tendency to interfere with one another, RFID transmission can be impeded by WLAN systems, by fluorescent lights, by electric motors, by cordless telephones, and a host of other devices.

The second major issue is cost. RFID remains an expensive solution, and although it offers compelling long-term advantages to the implementer, they come over time. Readers still cost approximately $1,000, although tags, which must be implemented in the hundreds or thousands if they are to generate a valid return, are $.30 to $.40 apiece. As deployment levels hit critical mass due to growing interest, the influence of early adopters, and the proliferation of legitimate standards, the

[1] Zebra's RFID Readiness Guidelines: Complying with RFID Tagging Mandates, an application whitepaper available at www.zebra.com/IS/white-papers.htm.

price for the devices will plummet in the relatively near future. For now, however, RFID remains expensive and that represents a significant barrier to adoption.

The third issue falls into the *be careful what you wish for, you might get it* category. A comprehensive RFID implementation is designed to collect data about the movement of assets as they move along the supply line—enormous volumes of it. RFID creates massive database entries, which ideally must be archived, filtered, analyzed, organized, and converted into comprehensive reports of useful information in real time. Today's data management applications are not equipped to handle the onslaught and will need serious work before they can handle it.

When RFID is deployed, it brings with it the requirement to modify business practices that have often been in place for a very long time. It is a well-documented fact that the vast majority of information portal implementations fail in their first few months of operation because they require that executives change their *modus operandi* in order to take advantage of them. The system should adapt to the user, not the other way around. RFID is no different; because it can mandate a dramatic change in the business processes into which it inserts itself—and it can include most of them—management and staff must be made aware of the coming changes early on to minimize the disruption of change that will inevitably result.

Finally, there is the question of expectations. Remember, RFID is designed to complement existing systems. It is also only as good as the data it creates and (perhaps more important) the analytical reports created from that data that yield useful business-oriented information. The implementation of an RFID-based business process overlay is more of a process than it is a project; it requires an ongoing effort to ensure that its deployment is successful and becomes embedded in the collective psyche of the corporation.

With all of these perceived caveats, it is important to manage the implementation properly to ensure that the perception of its success is a positive one. A thorough site survey should be performed prior to implementation to identify any physical, logical, or managerial constraints that could lead to a less-than-perfect implementation. For example, the survey should seek to identify all RF signal dead spots in the operational environment, as well as any particularly noisy areas that could interfere with RFID signal propagation. This survey should be conducted in close contact with the customer that will be using the system, because they and they alone will be most familiar with the track that tagged products will take as they make their way along the supply chain. Furthermore, it is critical that implementers understand every point along the way where the customer wishes to read tags as they pass by—along conveyor belts, in inside and outside doorways, in storage areas for theft control, on pallet jacks and forklifts, along hallways, inside vehicles, etc.

Equally important is the provisioning of necessary feeder plant. Although transponder tags require no connections other than being affixed to the product that they will track, readers are a different story. They require connections to power, to the antenna, and to the data network that will allow them to feed collected data back to the computer system that houses the RFID database. As the number of readers deployed in the operating area goes up, the complexity of the installed plant increases accordingly and must be planned for.

Similarly, the manner in which products make their way through the managed supply chain may dictate how the transponders are placed on products or on the containers that move them. For example, if the product is cases of canned soft drinks, it is critical that the tags to be read be placed on the same side of the container as the reader, due to the fact that the combination of metal and liquid is deadly as far as RF signal propagation is concerned.

Finally, it is critical to put into place a data management plan involving IT staff. As the data generated by the RFID system begins to accumulate, the underlying software must be configured to handle it —not just put it into appropriate data structures within an organized database, but also create reports and management information that can be applied to business challenges in rapid fashion. Remember "Garbage in, garbage out"? Well, this is a case of garbage in, garbage out, but with a serious attitude. Volumetrically, RFID will bury a corporation in data if it is not managed properly.

LOOKING AHEAD

With all that being said, what is the real future of RFID in the business marketplace? There is no question that its adoption will proceed apace and that it will achieve some levels of critical acceptance in the adoptive market. A collection of 500-pound gorillas have taken RFID by the scruff of the neck and are moving forward with enormous momentum to drive it to a successful level of adoption. Wal-Mart, the Department of Defense, Tesco, Albertsons, Target, and the *Food and Drug Administration* (FDA) have all placed their considerable influence behind the technology, and although they are moving forward slowly, they are moving forward. In early 2004, it was clear that Wal-Mart, the largest retailer in the world and the most influential player in the RFID game, was encountering some implementation challenges. The first was based on the pace of implementation. Most of the firm's suppliers simply weren't ready to move forward with RFID because they had not yet proven it in. Directly related to this was cost: The tags cost no less that $.25 each, and for the technology to be economically feasible, the price has to drop to approximately $.05, a decline that won't happen for some time. Some sectors may move more quickly than others, however. For example, in the pharmaceutical industry, where physical product security, tampering, and counterfeiting are serious concerns, RFID may become mainstream faster than it will in traditional retail because of the technology's ability to facilitate protection against these threats. And even more recently, the *Transportation Security Administration* (TSA) has announced that it intends to experiment with RFID-tagged boarding passes to improve flight security. They are already using the technology in Africa, where the *Safe Skies for Africa* initiative is underway.

A final concern is that of standards development. Standards are moving along at a stately pace, which translates into not fast enough. Unclear standards dramatically slowed the pace of implementation of 56 kbps modems back in the 1990s, and the same thing is now happening with RFID. Corporations are unwilling to make the required multimillion dollar investment in a technology that could change with the stroke of a collection of pens in Switzerland.

One standardization effort that shows significant promise is that of the EPC. EPCglobal, part of EAN International and the UCC, is close to completion of a standard for the preferred 96-bit EPC tags that most vendors are using. Even the Department of Defense has expressed its support for the emerging standard. And as the standard gets closer to broad acceptance, vendors have emerged with plans to offer low-cost transponders and readers. Alien Technologies, Symbol, and ThingMagic, for example, are all offering multifrequency tags that can be deployed for relatively low cost.

Table 4-1, adapted from Forrester Research, illustrates the most common deployments of RFID today.

THE RFID PLAYERS

New technologies and the companies that make products based on them are like minor injuries: Bang your elbow, and you will be painfully but regularly reminded of the existence of a body part that you were, prior to the injury, largely unaware of. The number of companies making RFID transponders, readers, and back office software are in the hundreds. In this section we present an overview of many of them. This list is far from conclusive, but it gives the reader a sense of the diversity in the sector.

MANUFACTURERS

This list is adapted from data provided by Transponder News, a highly useful source of RFID-related information. The site is kept

TABLE 4-1 Common Uses of RFID

OPERATING FREQUENCY	ADVANTAGES	LIABILITIES	APPLICATIONS
Low-Frequency: 125 kHz to 134 kHz	Widely deployed; broad global frequency deployment; metal interferes minimally	Read range limited to less than 1.5 meters	Animal tracking; container tracking; antitheft systems
High-Frequency: 13.56 MHz	Widely deployed; broad global frequency deployment; minimally affected by moisture	Read range limited to less than 1.5 meters; metal poses serious interference problem	Library asset tracking; access control; baggage tracking; retail product tracking
Ultra-High Frequency: 868 MHz to 928 MHz	Widely deployed; read range is significantly greater than other standards	Adversely affected by moisture; not licensed for use in Japan; adjacent tags cause detuning	Pallet, container tracking; vehicle tracking
Microwave: 2.45 GHz	Read range is significantly greater than other standards	Not widely deployed; complex implementation; not licensed in parts of Europe	Vehicle access control

current and provides portals to a wide variety of RFID companies and location services. Find them at http://transpondernews.com.

ACC Systems, Inc.: North American supplier of readers and terminals for 125 kHz and 13.56 MHz technology.

ACG Identification Technologies: Distributor of 125 kHz and 13.56 MHz transponders from multiple transponder manufacturers.

ADC SYSTEMS AB: Core industrial-grade reading modules for magnetically coupled RFID transponders with specific computer interfaces.

Advanced Interconnection Technology, Inc.: Manufacturer of coils for RFID antennas.

AEG Identifikationssysteme GmbH: High-volume manufacturing facilities for customized magnetic coupled transponders.

Agri Signal, Inc.: Manufacturer of various tagging systems for livestock herds.

Agricultural Technology Limited (ATL): Manufacturer of readers specialized for the rigorous requirements of dairy herd management.

Aleis International: Manufacturer of various tagging systems for livestock herds and feedlot operators.

Allsafe: Manufacturer of plastic cards for access control, identification, loyalty programs, and security applications.

AMB -i-t: Manufacturer of timing systems for motor racing.

Amskan: Manufacturer of industrial-grade RFID solutions.

AndroDat GmbH: Providers of mobile terminals, computers, and communication networks with integrated RFID readers.

Applied Wireless Identifications, Inc. (AWID): Manufacturer of readers for magnetically coupled transponders.

A.P.T. Smart Solutions: Supplier of RFID tagging systems.

Audemars Microtec: Producers of coils for magnetically coupled transponders.

Avid, Inc.: Manufacturers of tagging systems for pets and livestock.

Avonwood Developments Ltd.: Manufacturer of industrial-grade tagging systems.

Balogh: Manufacturers of RFID systems that operate at 125 kHz, 13.56 MHz, and 2.45 GHz, largely adapted for harsh industrial applications.

Bartec Dispensing Systems: Manufacturer of equipment for dispensing resins used in the encapsulation of RFID systems.

Baumer Ident AB: Providers of industrial-grade transponders.

Biomark: Manufacturer specializing in PIT Tag technology.

Brady RFID Systems: Providers of tags and readers.

CFG SA Microelectronics: Manufacturers of read/write heads for 125 kHz RFID transponders.

Cross Point b.v.: Providers of EAS systems and RFID systems.

Datamars: Manufacturer of livestock tracking-oriented RFID products.

Deister Electronic GmbH: Suppliers of magnetically coupled transponders and readers for industrial, access, and sport solutions.

Dorian Industrier Pty Ltd.: Manufacturers of tagging systems for horse racing and timing systems for auto racing and horse racing.

DTE Automation GmbH: Providers of RFID computers and wands for magnetically coupled transponders.

ELPAS Ltd.: Designers of RFID technology that is integrated into applications for smart hospitals and intelligent buildings.

EM Microelectronic-Marin SA: Manufacturers of integrated circuits and modules for transponders.

Escort Memory Systems: Manufacturers of tagging systems for a wide range of industrial applications.

Euchner & Co: RFID-based tool and pallet coding devices.

eXI Wireless, Inc.: Manufacturers of a transponder system designed to monitor patients and infants in hospitals and to track assets and personnel.

Extel srl: RFID access control system for personnel and vehicles.

Gantner Electronic GmbH: Provider of RFID components for access control, industrial applications, and sports solutions.

HERMOS Informatik GmbH: Manufacturers of tracking systems for wafer cartridges in semiconductor foundries.

Hi-G-Tek Ltd.: Transponders and readers to monitor sealed shipping containers.

HID Corporation: Manufacturers and suppliers of 125 kHz and 13.56 MHz transponders and readers for access control solutions.

HOTRACO MicroID BV: Manufacturers of tagging systems for livestock monitoring.

IB Technology Ltd.: Suppliers of a single chip reader module for 125 kHz transponders.

Identec Ltd(UK): Suppliers and developers of Cryptag and Cryptag Census transponders.

Identec Solutions, Inc.: Suppliers and developers of RFID transponders that can monitor environmental conditions although tags are in transit.

Idesco: Providers of RFID products for factory automation, access control, and ticketing applications.

Infineon Technologies AG: Manufacturers of a broad range of RFID components, but they are particularly well known for manufacturing the first RFID products with strong security algorithms that enable the Chip Sharing Approach. Their solution can also be used as a powerful tool for anticounterfeiting and data protection.

Intersoft: Manufacturers of magnetically coupled tags and readers.

LEGIC Identsystems: Manufacturers of 13.56 MHz contactless RFID and smartcard systems.

Magnatec Technologie GmbH: An intriguing application that combines RFID and GSM technologies for monitoring security guards' movements.

Matrics Ltd.: Suppliers of integrated chips for 125 kHz, HF, and UHF transponders.

Melexis: Manufacturers of integrated circuits for sensors and magnetic coupled transponders.

Message Xpress: Provider of RFID tracking systems for monitoring vehicle fleet movements.

METGET AB: Manufacturers of transponder coils and packaging for custom transponders.

Microchip Technology: Manufacturer of passive contactless programmable 125 kHz RFID devices.

Nedap: Manufacturer of vehicle and driver identification systems.

NWK Technology: Manufacturers of readers for magnetically coupled transponders.

OmniTek: Provider of reader systems for 125 kHz and 13.56 MHz access control applications.

PAV CARD GmbH: Manufacturers of 13.56 MHz contactless RFID and smartcard systems.

Pepperl + Fuchs, Inc.: Industrial transponder systems for factory automation.

Quelis Id Systems: Manufacturers of transponders for 125 kHz and 13.56 MHz technologies.

RF Technologies: Provider of RFID solutions for monitoring patients in hospitals.

RFID Systems Corporation: Specialize in integrated RFID technology solutions.

RJI Solutions, Inc.: Providers of RFID solutions for the laundry and textile rental industries.

Scemtec: Manufacturers of reading and writing systems for magnetic coupled transponder systems.

Sedco Engineering: Tagging systems for mining operations.

Siemens AG: Transponder systems for logistics and manufacturing.

SOKYMAT: World's largest supplier of RFID transponders and related devices.

STMicroelectronics: Manufacturers of contactless smart card chips operating at 13.56 MHz.

Stoval Technologies Pte Ltd.: Manufacturers of smart cards and readers.

Tadiran Telematics Ltd.: FASTPASS electronic vehicle toll and traffic management systems.

Tagmaster AB: Manufacturer of microwave tags.

Tagsys (Gemplus Tag): Providers of smart labels and RFID systems.

TEK Industries, Inc.: Manufacturers of combined barcode and RFID scanners.

Telesensomatik: Manufacturer of contactless systems.

Texas Instruments: Manufacturers of magnetic and electric coupled transponders and readers.

Traffic Supervision Systems A/S: RFID systems for monitoring transportation systems.

Transcore: Tagging systems for rail rolling stock, toll roads, and parking systems.

Trolley Scan (Pty) Ltd.: Supplier of low-power passive UHF ISO card-sized transponders and readers, and EcoTag UHF RFID transponder systems.

Trovan: Manufacturer of electronic identification systems.

Xmark: Manufacturers of monitoring systems for elderly patients and babies in healthcare facilities.

Zebra Technologies: Provider of innovative and reliable barcode and specialty label printing solutions to businesses and governments in 90 countries around the world.

FINAL THOUGHTS

One hot summer evening in the North Atlantic, 250 miles from the ship's destination at the Port of New York and New Jersey, a massive container ship carrying an almost incomprehensible 4,000 containers finds itself surrounded by a flotilla of Coast Guard vessels that order it to stop. The vessel is boarded by an armed search party, which goes directly to a particular container in the maze of stacked boxes on the ship. Opening it, they find drugs, or people, or worse. Any threat carried in the container has been stopped before entering port through the use of advanced monitoring technologies. One of them is RFID.

The world's container port system is one of the most magically complex and important components of the global economy. It is powerfully robust, and yet is also dangerously fragile. Millions of containers a year move in and out of container ports, yet customs officials are hard-pressed to closely inspect more than two to three percent of them. The job is simply too large, too daunting. It is a job, however, that desperately needs doing. Consider the following statistics. From the time a container is loaded onto a ship until it is taken off the ship at its final destination, it has, on average, passed through seven intermediate port facilities, offering seven opportunities for tampering. Multiply that by the number of ports, the number of ships, the number of containers, and the number of customs agents, and you begin to see the nature of the challenge.

A retail executive I interviewed for this book was quite blunt about the inherent danger on economic terms. "My firm has 1,500 stores in 140 countries and buys products from 18,000 distributors around the world. Our global logistics systems are so good that we really can do just-in-time inventory control—and we're very, very good at it. It allows us to keep on-hand inventories extremely low, thus avoiding the cost of warehousing. In fact, our most inventory-heavy stores only have six days worth of product on hand.

Now, consider the following scenario: A terrorism-related event causes the government to respond in such a way that they feel compelled to shut down the global shipping system; nothing comes into port until they deem it safe. In a maximum of six days, I'm out of business: 1,500 stores are empty. And it isn't just my company—it's every firm in the country that relies on imported stock. And well over 90 percent of all imported stock finds itself at one time or another in a shipping container. So this system may be powerful and robust, but it is also dangerously vulnerable. We need technology solutions that will allow us to monitor and inspect all containers and to know if and when they have been tampered with so that we can preemptively manage the problem."

This is a serious and very real scenario. If the problem described were to occur, and if shipping were to be stopped, a major disruption of international commerce would take place. Governments need to take action *now* to identify and put into place systems and technology-based solutions that will facilitate continuance of trade in case of a disruptive event. Waiting until it happens is not the answer; proactive problem management is needed.

There are companies building solutions to this challenge right now. Nevada Technology Corporation, for example, has designed a highly innovative smart container that can detect an intrusion and notify the appropriate authorities in real time that a container has been opened. Their system can also link into accelerometers, light detectors, sound detectors, temperature sensors, nuclear detectors, and a variety of other sensor devices and can report the information detected by those devices in a number of ways.

The point of all this is that global shipping, logistics management, and supply chain management are intertwined components of global commerce. The massive container vessels such as the *Cornelius Maersk* shown in Figure 4-5 can carry massive volumes of cargo and represent the heart and soul of global commerce.

RFID is positioned to play a key role in port and container security. For example, companies like Savi Technology, EJ Brooks, and Hi-G-Tek manufacture electronic RFID-based seals for shipping containers. These seals detect when they are broken and have the ability to send a notification of the fact to a collection point on the ship, which can in turn transmit the information via satellite to a shore-based collection point. The collection point provides access to behind-the-scenes analysis applications that have the ability to correlate the collected data through sophisticated data mining techniques and craft an appropriate response, in real time, simultaneously notifying the appropriate authorities to take whatever action they deem necessary.

FIGURE 4-5 A massive vessel such as the *Cornelius Maersk,* shown here in Singapore Harbor, can carry an unimaginable amount of cargo. These vessels are the heart of global commerce. *(Photo courtesy Maersk Shipping.)*

This is what RFID is all about. As we said earlier, it is an asset tracking technology, no more and no less. The stakes, however, are high, and its deployment will help to manage them on a global basis.

APPENDIX A

COMMON INDUSTRY ACRONYMS

AAL	ATM Adaptation Layer
AARP	AppleTalk Address Resolution Protocol
ABM	Asynchronous Balanced Mode
ABR	Available Bit Rate
AC	Alternating Current
ACD	Automatic Call Distribution
ACELP	Algebraic Code-Excited Linear Prediction
ACF	Advanced Communication Function
ACK	Acknowledgment
ACM	Address Complete Message
ACSE	Association Control Service Element
ACTLU	Activate Logical Unit
ACTPU	Activate Physical Unit
ADCCP	Advanced Data Communications Control Procedures
ADM	Add/Drop Multiplexer
ADPCM	Adaptive Differential Pulse Code Modulation
ADSL	Asymmetric Digital Subscriber Line
AFI	Application Family Identifier (RFID)
AFI	Authority and Format Identifier
AI	Application Identifier
AIN	Advanced Intelligent Network
AIS	Alarm Indication Signal
ALU	Arithmetic Logic Unit
AM	Administrative Module (Lucent 5ESS)

AM	Amplitude Modulation
AMI	Alternate Mark Inversion
AMP	Administrative Module Processor
AMPS	Advanced Mobile Phone System
ANI	Automatic Number Identification (SS7)
ANSI	American National Standards Institute
APD	Avalanche Photodiode
API	Application Programming Interface
APPC	Advanced Program-to-Program Communication
APPN	Advanced Peer-to-Peer Networking
APS	Automatic Protection Switching
ARE	All Routes Explorer (Source Route Bridging)
ARM	Asynchronous Response Mode
ARP	Address Resolution Protocol (IETF)
ARPA	Advanced Research Projects Agency
ARPANET	Advanced Research Projects Agency Network
ARPU	Average Revenue per User
ARQ	Automatic Repeat Request
ASCII	American Standard Code for Information Interchange
ASI	Alternate Space Inversion
ASIC	Application Specific Integrated Circuit
ASK	Amplitude Shift Keying
ASN	Abstract Syntax Notation
ASP	Application Service Provider
AT&T	American Telephone and Telegraph
ATDM	Asynchronous Time Division Multiplexing
ATM	Asynchronous Transfer Mode
ATM	Automatic Teller Machine
ATMF	ATM Forum
ATQA	Answer to Request A (RFID)
ATQB	Answer to Request B (RFID)
ATS	Answer to Select (RFID)
ATTRIB	Attribute (RFID)
AU	Administrative Unit (SDH)
AUG	Administrative Unit Group (SDH)
AWG	American Wire Gauge

B-ICI	Broadband Intercarrier Interface
B-ISDN	Broadband Integrated Services Digital Network
B2B	Business-to-Business
B2C	Business-to-Consumer
B8ZS	Binary 8 Zero Substitution
BANCS	Bell Administrative Network Communications System
BBN	Bolt, Beranak, and Newman
BBS	Bulletin Board Service
Bc	Committed Burst Size
BCC	Block Check Character
BCC	Blocked Calls Cleared
BCD	Blocked Calls Delayed
BCDIC	Binary Coded Decimal Interchange Code
Be	Excess Burst Size
BECN	Backward Explicit Congestion Notification
BER	Bit Error Rate
BERT	Bit Error Rate Test
BGP	Border Gateway Protocol (IETF)
BIB	Backward Indicator Bit (SS7)
BIOS	Basic Input/Output System
BIP	Bit Interleaved Parity
BISYNC	Binary Synchronous Communications Protocol
BITNET	Because It's Time Network
BITS	Building Integrated Timing Supply
BLSR	Bidirectional Line Switched Ring
BOC	Bell Operating Company
BPRZ	Bipolar Return to Zero
Bps	Bits per Second
BRI	Basic Rate Interface
BRITE	Basic Rate Interface Transmission Equipment
BSC	Binary Synchronous Communications
BSN	Backward Sequence Number (SS7)
BSRF	Bell System Reference Frequency
BTAM	Basic Telecommunications Access Method
BUS	Broadcast Unknown Server
C/R	Command/Response
CAD	Computer-Aided Design

CAE	Computer-Aided Engineering
CAGR	Compound Annual Growth Rate
CAM	Computer-Aided Manufacturing
CAP	Carrierless Amplitude/Phase modulation
CAP	Competitive Access Provider
CAPEX	Capital Expenditure
CAPEX	Capital Expense
CARICOM	Caribbean Community and Common Market
CASE	Common Application Service Element
CASE	Computer-Aided Software Engineering
CASPIAN	Consumers Against Privacy Invasion and Numbering (RFID)
CAT	Computer-Aided Tomography
CATIA	Computer-Assisted Three-dimensional Interactive Application
CATV	Community Antenna Television
CBEMA	Computer and Business Equipment Manufacturers Association
CBR	Constant Bit Rate
CBT	Computer-Based Training
CC	Cluster Controller
CCIR	International Radio Consultative Committee
CCIS	Common Channel Interoffice Signaling
CCITT	International Telegraph and Telephone Consultative Committee
CCS	Common Channel Signaling
CCS	Hundred Call Seconds per Hour
CD	Collision Detection
CD	Compact Disc
CD-ROM	Compact Disc-Read Only Memory
CDC	Control Data Corporation
CDMA	Code Division Multiple Access
CDPD	Cellular Digital Packet Data
CDVT	Cell Delay Variation Tolerance
CEI	Comparably Efficient Interconnection
CEPT	Conference of European Postal and Telecommunications Administrations
CERN	European Council for Nuclear Research

CERT	Computer Emergency Response Team
CES	Circuit Emulation Service
CEV	Controlled Environmental Vault
CGI	Common Gateway Interface (Internet)
CHAP	Challenge Handshake Authentication Protocol
CHL	Chain Home Low radar
CICS	Customer Information Control System
CICS/VS	Customer Information Control System/Virtual Storage
CID	Card Identifier (RFID)
CIDR	Classless Interdomain Routing (IETF)
CIF	Cells In Frames
CIR	Committed Information Rate
CISC	Complex Instruction Set Computer
CIX	Commercial Internet Exchange
CLASS	Custom Local Area Signaling Services (Bellcore)
CLEC	Competitive Local Exchange Carrier
CLLM	Consolidated Link Layer Management
CLNP	Connectionless Network Protocol
CLNS	Connectionless Network Service
CLP	Cell Loss Priority
CM	Communications Module (Lucent 5ESS)
CMIP	Common Management Information Protocol
CMISE	Common Management Information Service Element
CMOL	CMIP Over LLC
CMOS	Complementary Metal Oxide Semiconductor
CMOT	CMIP Over TCP/IP
CMP	Communications Module Processor
CNE	Certified NetWare Engineer
CNM	Customer Network Management
CNR	Carrier-to-Noise Ratio
CO	Central Office
CoCOM	Coordinating Committee on Export Controls
CODEC	Coder-Decoder
COMC	Communications Controller
CONS	Connection-Oriented Network Service

CORBA	Common Object Request Brokered Architecture
COS	Class of Service (APPN)
COS	Corporation for Open Systems
CPE	Customer Premises Equipment
CPU	Central Processing Unit
CRC	Cyclic Redundancy Check
CRM	Customer Relationship Management
CRT	Cathode Ray Tube
CRV	Call Reference Value
CS	Convergence Sublayer
CSA	Carrier Serving Area
CSMA	Carrier Sense Multiple Access
CSMA/CA	Carrier Sense Multiple Access with Collision Avoidance
CSMA/CD	Carrier Sense Multiple Access with Collision Detection
CSU	Channel Service Unit
CTI	Computer Telephony Integration
CTIA	Cellular Telecommunications Industry Association
CTS	Clear To Send
CU	Control Unit
CVSD	Continuously Variable Slope Delta modulation
CWDM	Coarse Wavelength Division Multiplexing
D/A	Digital-to-Analog
DA	Destination Address
DAC	Dual Attachment Concentrator (FDDI)
DACS	Digital Access and Cross-connect System
DARPA	Defense Advanced Research Projects Agency
DAS	Direct Attached Storage
DAS	Dual Attachment Station (FDDI)
DASD	Direct Access Storage Device
DB	Decibel
DBS	Direct Broadcast Satellite
DC	Direct Current
DCC	Data Communications Channel (SONET)
DCE	Data Circuit-terminating Equipment

DCN	Data Communications Network
DCS	Digital Cross-connect System
DCT	Discrete Cosine Transform
DDCMP	Digital Data Communications Management Protocol (DNA)
DDD	Direct Distance Dialing
DDP	Datagram Delivery Protocol
DDS	DATAPHONE Digital Service (Sometimes Digital Data Service)
DE	Discard Eligibility (LAPF)
DECT	Digital European Cordless Telephone
DES	Data Encryption Standard (NIST)
DID	Direct Inward Dialing
DIP	Dual Inline Package
DLC	Digital Loop Carrier
DLCI	Data Link Connection Identifier
DLE	Data Link Escape
DLSw	Data Link Switching
DM	Data Mining
DM	Delta Modulation
DM	Disconnected Mode
DMA	Direct Memory Access (computers)
DMAC	Direct Memory Access Control
DME	Distributed Management Environment
DMS	Digital Multiplex Switch
DMT	Discrete Multitone
DNA	Digital Network Architecture
DNIC	Data Network Identification Code (X.121)
DNIS	Dialed Number Identification Service
DNS	Domain Name Service
DNS	Domain Name System (IETF)
DOD	Department of Defense
DOD	Direct Outward Dialing
DOJ	Department of Justice
DOV	Data Over Voice
DPSK	Differential Phase Shift Keying
DQDB	Distributed Queue Dual Bus
DR	Data Rate Send (RFID)

DRAM	Dynamic Random Access Memory
DS	Data Rate Send (RFID)
DSAP	Destination Service Access Point
DSF	Dispersion-Shifted Fiber
DSI	Digital Speech Interpolation
DSL	Digital Subscriber Line
DSLAM	Digital Subscriber Line Access Multiplexer
DSP	Digital Signal Processing
DSR	Data Set Ready
DSS	Digital Satellite System
DSSS	Digital Subscriber Signaling System
DSSS	Direct Sequence Spread Spectrum
DSU	Data Service Unit
DTE	Data Terminal Equipment
DTMF	Dual Tone Multifrequency
DTR	Data Terminal Ready
DVRN	Dense Virtual Routed Networking (Crescent)
DWDM	Dense Wavelength Division Multiplexing
DXI	Data Exchange Interface
E/O	Electrical-to-Optical
EAN	European Article Numbering System
EBCDIC	Extended Binary Coded Decimal Interchange Code
EBITDA	Earnings before Interest, Tax, Depreciation, and Amortization
ECMA	European Computer Manufacturer Association
ECN	Explicit Congestion Notification
ECSA	Exchange Carriers Standards Association
EDFA	Erbium-Doped Fiber Amplifier
EDI	Electronic Data Interchange
EDIBANX	EDI Bank Alliance Network Exchange
EDIFACT	Electronic Data Interchange For Administration, Commerce, and Trade (ANSI)
EFCI	Explicit Forward Congestion Indicator
EFTA	European Free Trade Association
EGP	Exterior Gateway Protocol (IETF)
EIA	Electronics Industry Association
EIGRP	Enhanced Interior Gateway Routing Protocol

EIR	Excess Information Rate
EMBARC	Electronic Mail Broadcast to a Roaming Computer
EMI	Electromagnetic Interference
EMS	Element Management System
EN	End Node
ENIAC	Electronic Numerical Integrator and Computer
EO	End Office
EOC	Embedded Operations Channel (SONET)
EOT	End of Transmission (BISYNC)
EPC	Electronic Product Code
EPROM	Erasable Programmable Read Only Memory
EPS	Earnings per Share
ERP	Enterprise Resource Planning
ESCON	Enterprise System Connection (IBM)
ESF	Extended Superframe Format
ESOP	Employee Stock Ownership Plan
ESP	Enhanced Service Provider
ESS	Electronic Switching System
ETSI	European Telecommunications Standards Institute
ETX	End of Text (BISYNC)
EVA	Economic Value Added
EWOS	European Workshop for Open Systems
FACTR	Fujitsu Access and Transport System
FAQ	Frequently Asked Questions
FASB	Financial Accounting Standards Board
FAT	File Allocation Table
FCF	Free Cash Flow
FCS	Frame Check Sequence
FDA	Food and Drug Administration
FDD	Frequency Division Duplex
FDDI	Fiber Distributed Data Interface
FDM	Frequency Division Multiplexing
FDMA	Frequency Division Multiple Access
FDX	Full-Duplex
FEBE	Far End Block Error (SONET)
FEC	Forward Error Correction

FEC	Forward Equivalence Class
FECN	Forward Explicit Congestion Notification
FEP	Front-End Processor
FERF	Far End Receive Failure (SONET)
FET	Field Effect Transistor
FHSS	Frequency Hopping Spread Spectrum
FIB	Forward Indicator Bit (SS7)
FIFO	First In First Out
FITL	Fiber In The Loop
FLAG	Fiber Ling Across the Globe
FM	Frequency Modulation
FOIRL	Fiber Optic Inter-Repeater Link
FPGA	Field Programmable Gate Array
FR	Frame Relay
FRAD	Frame Relay Access Device
FRBS	Frame Relay Bearer Service
FSDI	Frame Size Device Integer (RFID)
FSK	Frequency Shift Keying
FSN	Forward Sequence Number (SS7)
FTAM	File Transfer, Access, and Management
FTP	File Transfer Protocol (IETF)
FTTC	Fiber to the Curb
FTTH	Fiber to the Home
FUNI	Frame User-to-Network Interface
FWI	Frame Waiting Integer (RFID)
FWM	Four Wave Mixing
GAAP	Generally Accepted Accounting Principles
GATT	General Agreement on Tariffs and Trade
GbE	Gigabit Ethernet
Gbps	Gigabits per second (billion bits per second)
GDMO	Guidelines for the Development of Managed Objects
GDP	Gross Domestic Product
GEOS	Geosynchronous Earth Orbit Satellites
GFC	Generic Flow Control (ATM)
GFI	General Format Identifier (X.25)
GFP	Generic Framing Procedure

GFP-F	Generic Framing Procedure-Frame-Based
GFP-X	Generic Framing Procedure-Transparent
GMPLS	Generalized MPLS
GOSIP	Government Open Systems Interconnection Profile
GPS	Global Positioning System
GRIN	Graded Index (fiber)
GSM	Global System for Mobile Communications
GTIN	Global Trade Item Number
GUI	Graphical User Interface
HDB3	High Density, Bipolar 3 (E-Carrier)
HDLC	High-level Data Link Control
HDSL	High-bit-rate Digital Subscriber Line
HDTV	High Definition Television
HDX	Half-Duplex
HEC	Header Error Control (ATM)
HFC	Hybrid Fiber/Coax
HFS	Hierarchical File Storage
HIPAA	Health Insurance Portability and Accountability Act
HLR	Home Location Register
HPPI	High Performance Parallel Interface
HSSI	High-Speed Serial Interface (ANSI)
HTML	Hypertext Markup Language
HTTP	Hypertext Transfer Protocol (IETF)
HTU	HDSL Transmission Unit
I	Intrapictures
IAB	Internet Architecture Board (formerly Internet Activities Board)
IACS	Integrated Access and Cross-connect System
IAD	Integrated Access Device
IAM	Initial Address Message (SS7)
IANA	Internet Address Naming Authority
ICMP	Internet Control Message Protocol (IETF)
IDP	Internet Datagram Protocol
IEC	Interexchange Carrier (also IXC)
IEC	International Electrotechnical Commission

IEEE	Institute of Electrical and Electronics Engineers
IETF	Internet Engineering Task Force
IFRB	International Frequency Registration Board
IGP	Interior Gateway Protocol (IETF)
IGRP	Interior Gateway Routing Protocol
ILEC	Incumbent Local Exchange Carrier
IM	Instant Messenger (AOL)
IML	Initial Microcode Load
IMP	Interface Message Processor (ARPANET)
IMS	Information Management System
InARP	Inverse Address Resolution Protocol (IETF)
InATMARP	Inverse ATMARP
INMARSAT	International Maritime Satellite Organization
INP	Internet Nodal Processor
InterNIC	Internet Network Information Center
IP	Intellectual Property
IP	Internet Protocol (IETF)
IPO	Initial Product Offer
IPX	Internetwork Packet Exchange (NetWare)
IRU	Indefeasible Rights of Use
IS	Information Systems
ISDN	Integrated Services Digital Network
ISO	Information Systems Organization
ISO	International Organization for Standardization
ISOC	Internet Society
ISP	Internet Service Provider
ISUP	ISDN User Part (SS7)
IT	Information Technology
ITU	International Telecommunication Union
ITU-R	International Telecommunication Union-Radio Communication Sector
IVD	Inside Vapor Deposition
IVR	Interactive Voice Response
IXC	Interexchange Carrier
JAN	Japanese Article Numbering System
JEPI	Joint Electronic Paynets Initiative
JES	Job Entry System

JIT	Just in Time
JPEG	Joint Photographic Experts Group
JTC	Joint Technical Committee
kB	Kilobytes
kbps	Kilobits per second (thousand bits per second)
KLTN	Potassium Lithium Tantalate Niobate
KM	Knowledge Management
LAN	Local Area Network
LANE	LAN Emulation
LAP	Link Access Procedure (X.25)
LAPB	Link Access Procedure Balanced (X.25)
LAPD	Link Access Procedure for the D-Channel
LAPF	Link Access Procedure to Frame Mode Bearer Services
LAPF-Core	Core Aspects of the Link Access Procedure to Frame Mode Bearer Services
LAPM	Link Access Procedure for Modems
LAPX	Link Access Procedure half-duplex
LASER	Light Amplification by the Stimulated Emission of Radiation
LATA	Local Access and Transport Area
LCD	Liquid Crystal Display
LCGN	Logical Channel Group Number
LCM	Line Concentrator Module
LCN	Local Communications Network
LD	Laser Diode
LDAP	Lightweight Directory Access Protocol (X.500)
LEAF®	Large Effective Area Fiber® (Corning product)
LEC	Local Exchange Carrier
LED	Light Emitting Diode
LENS	Lightwave Efficient Network Solution (Centerpoint)
LEOS	Low Earth Orbit Satellites
LER	Label Edge Router
LI	Length Indicator
LIDB	Line Information Database
LIFO	Last In First Out
LIS	Logical IP Subnet

LLC	Logical Link Control
LMDS	Local Multipoint Distribution System
LMI	Local Management Interface
LMOS	Loop Maintenance Operations System
LORAN	Long-range Radio Navigation
LPC	Linear Predictive Coding
LPP	Lightweight Presentation Protocol
LRC	Longitudinal Redundancy Check (BISYNC)
LS	Link State
LSI	Large Scale Integration
LSP	Label Switched Path
LSR	Label Switched Router
LU	Line Unit
LU	Logical Unit (SNA)
MAC	Media Access Control
MAN	Metropolitan Area Network
MAP	Manufacturing Automation Protocol
MAU	Medium Attachment Unit (Ethernet)
MAU	Multistation Access Unit (Token Ring)
MB	Megabytes
MBA™	Metro Business Access™ (Ocular)
Mbps	Megabits per second (million bits per second)
MD	Message Digest (MD2, MD4, MD5) (IETF)
MDF	Main Distribution Frame
MDU	Multi-Dwelling Unit
MEMS	Micro Electrical Mechanical System
MF	Multifrequency
MFJ	Modified Final Judgment
MHS	Message Handling System (X.400)
MIB	Management Information Base
MIC	Medium Interface Connector (FDDI)
MIME	Multipurpose Internet Mail Extensions (IETF)
MIPS	Millions of Instructions Per Second
MIS	Management Information Systems
MITI	Ministry of International Trade and Industry (Japan)
MITS	Micro Instrumentation and Telemetry Systems

ML-PPP	Multilink Point-to-Point Protocol
MMDS	Multichannel, Multipoint Distribution System
MMF	Multimode Fiber
MNP	Microcom Networking Protocol
MON	Metropolitan Optical Network
MoU	Memorandum of Understanding
MP	Multilink PPP
MPλS	Multiprotocol Lambda Switching
MPEG	Motion Picture Experts Group
MPLS	Multiprotocol Label Switching
MPOA	Multiprotocol Over ATM
MRI	Magnetic Resonance Imaging
MSB	Most Significant Bit
MSC	Mobile Switching Center
MSO	Mobile Switching Office
MSPP	Multi-Service Provisioning Platform
MSVC	Meta-signaling Virtual Channel
MTA	Major Trading Area
MTBF	Mean Time Between Failure
MTP	Message Transfer Part (SS7)
MTSO	Mobile Telephone Switching Office
MTTR	Mean Time to Repair
MTU	Maximum Transmission Unit
MTU	Multi-Tenant Unit
MVNO	Mobile Virtual Network Operator
MVS	Multiple Virtual Storage
NAD	Node Address (RFID)
NAFTA	North American Free Trade Agreement
NAK	Negative Acknowledgment (BISYNC, DDCMP)
NAP	Network Access Point (Internet)
NARUC	National Association of Regulatory Utility Commissioners
NAS	Network Attached Storage
NASA	National Aeronautics and Space Administration
NASDAQ	National Association of Securities Dealers Automated Quotations

NATA	North American Telecommunications Association
NATO	North Atlantic Treaty Organization
NAU	Network Accessible Unit
NCP	Network Control Program
NCSA	National Center for Supercomputer Applications
NCTA	National Cable Television Association
NDIS	Network Driver Interface Specifications
NDSF	Non-Dispersion-Shifted Fiber
NetBEUI	NetBIOS Extended User Interface
NetBIOS	Network Basic Input/Output System
NFS	Network File System (Sun)
NIC	Network Interface Card
NII	National Information Infrastructure
NIST	National Institute of Standards and Technology (formerly NBS)
NIU	Network Interface Unit
NLPID	Network Layer Protocol Identifier
NLSP	NetWare Link Services Protocol
NM	Network Module
Nm	Nanometer
NMC	Network Management Center
NMS	Network Management System
NMT	Nordic Mobile Telephone
NMVT	Network Management Vector Transport protocol
NNI	Network Node Interface
NNI	Network-to-Network Interface
NOC	Network Operations Center
NOCC	Network Operations Control Center
NOPAT	Net Operating Profit After Tax
NOS	Network Operating System
NPA	Numbering Plan Area
NREN	National Research and Education Network
NRZ	Non-Return to Zero
NRZI	Non-Return to Zero Inverted
NSA	National Security Agency

NSAP	Network Service Access Point
NSAPA	Network Service Access Point Address
NSF	National Science Foundation
NTSC	National Television Systems Committee
NTT	Nippon Telephone and Telegraph
NVB	Number of Valid Bits (RFID)
NVOD	Near Video on Demand
NZDSF	Non-Zero Dispersion-Shifted Fiber
O-E-O	Optical-Electrical-Optical
OADM	Optical Add-Drop Multiplexer
OAM	Operations, Administration, and Maintenance
OAM&P	Operations, Administration, Maintenance, and Provisioning
OAN	Optical Area Network
OBS	Optical Burst Switching
OC	Optical Carrier
OEM	Original Equipment Manufacturer
OLS	Optical Line System (Lucent)
OMAP	Operations, Maintenance, and Administration Part (SS7)
ONA	Open Network Architecture
ONS	Object Name Service
ONU	Optical Network Unit
OOF	Out of Frame
OPEX	Operating Expense
OS	Operating System
OSF	Open Software Foundation
OSI	Open Systems Interconnection (ISO, ITU-T)
OSI RM	Open Systems Interconnection Reference Model
OSPF	Open Shortest Path First (IETF)
OSS	Operation Support Systems
OTDM	Optical Time Division Multiplexing
OTDR	Optical Time-Domain Reflectometer
OUI	Organizationally Unique Identifier (SNAP)
OVD	Outside Vapor Deposition
OXC	Optical Cross-Connect
P/F	Poll/Final (HDLC)

PAD	Packet Assembler/Disassembler (X.25)
PAL	Phase Alternate Line
PAM	Pulse Amplitude Modulation
PANS	Pretty Amazing New Stuff
PBX	Private Branch Exchange
PCB	Protocol Control Byte (RFID)
PCI	Peripheral Component Interface
PCM	Pulse Code Modulation
PCMCIA	Personal Computer Memory Card International Association
PCN	Personal Communications Network
PCS	Personal Communications Services
PDA	Personal Digital Assistant
PDH	Plesiochronous Digital Hierarchy
PDU	Protocol Data Unit
PIN	Positive-Intrinsic-Negative
PING	Packet Internet Groper (TCP/IP)
PKC	Public Key Cryptography
PLCP	Physical Layer Convergence Protocol
PLP	Packet Layer Protocol (X.25)
PM	Phase Modulation
PMD	Physical Medium Dependent (FDDI)
PML	Physical Markup Language
PNNI	Private Network Node Interface (ATM)
PON	Passive Optical Networking
POP	Point of Presence
POSIT	Profiles for Open Systems Interworking Technologies
POSIX	Portable Operating System Interface for UNIX
POTS	Plain Old Telephone Service
PPM	Pulse Position Modulation
PPP	Point-to-Point Protocol (IETF)
PPS	Protocol Parameter Selection (RFID)
PRC	Primary Reference Clock
PRI	Primary Rate Interface
PROFS	Professional Office System
PROM	Programmable Read Only Memory

PSDN	Packet Switched Data Network
PSK	Phase Shift Keying (RFID)
PSPDN	Packet Switched Public Data Network
PSTN	Public Switched Telephone Network
PTI	Payload Type Identifier (ATM)
PTT	Post, Telephone, and Telegraph
PU	Physical Unit (SNA)
PUC	Public Utility Commission
PUPI	Pseudo-Unique PICC Identifier
PVC	Permanent Virtual Circuit
Q-bit	Qualified data bit (X.25)
QAM	Quadrature Amplitude Modulation
QLLC	Qualified Logical Link Control (SNA)
QoS	Quality of Service
QPSK	Quadrature Phase Shift Keying
QPSX	Queued Packet Synchronous Exchange
R&D	Research & Development
RADAR	Radio Detection and Ranging
RADSL	Rate Adaptive Digital Subscriber Line
RAID	Redundant Array of Inexpensive Disks
RAM	Random Access Memory
RARP	Reverse Address Resolution Protocol (IETF)
RAS	Remote Access Server
RATS	Request for Answer to Select (RFID)
RBOC	Regional Bell Operating Company
READ_DATA	Read Data from Transponder (RFID)
REQA	Request A (RFID)
REQB	Request B (RFID)
REQUEST_SNR	Request Serial Number (RFID)
RF	Radio Frequency
RFC	Request For Comments (IETF)
RFH	Remote Frame Handler (ISDN)
RFI	Radio Frequency Interference
RFID	Radio Frequency Identification
RFP	Request for Proposal
RFQ	Request for Quote

RFx	Request for X, where "X" can be Proposal, Quote, Information, Comment, etc.
RHC	Regional Holding Company
RHK	Ryan, Hankin, and Kent (Consultancy)
RIP	Routing Information Protocol (IETF)
RISC	Reduced Instruction Set Computer
RJE	Remote Job Entry
RNR	Receive Not Ready (HDLC)
RO-RO	Roll-On Roll-Off
ROA	Return on Assets
ROE	Return on Equity
ROI	Return on Investment
ROM	Read-Only Memory
ROSE	Remote Operation Service Element
RPC	Remote Procedure Call
RPR	Resilient Packet Ring
RR	Receive Ready (HDLC)
RSA	Rivest, Shamir, and Aleman
RTS	Request To Send (EIA-232-E)
S/DMS	SONET/Digital Multiplex System
S/N	Signal-to-Noise Ratio
S-HTTP	Secure HTTP (IETF)
SAA	Systems Application Architecture (IBM)
SAAL	Signaling ATM Adaptation Layer (ATM)
SABM	Set Asynchronous Balanced Mode (HDLC)
SABME	Set Asynchronous Balanced Mode Extended (HDLC)
SAC	Single Attachment Concentrator (FDDI)
SAK	Select Acknowledge (RFID)
SAN	Storage Area Network
SAP	Service Access Point (generic)
SAPI	Service Access Point Identifier (LAPD)
SAR	Segmentation and Reassembly (ATM)
SAS	Single Attachment Station (FDDI)
SASE	Specific Applications Service Element (subset of CASE, Application Layer)
SATAN	System Administrator Tool for Analyzing Networks

SBS	Stimulated Brillouin Scattering
SCCP	Signaling Connection Control Point (SS7)
SCM	Supply Chain Management
SCP	Service Control Point (SS7)
SCREAM™	Scalable Control of a Rearrangeable Extensible Array of Mirrors (Calient)
SCSI	Small Computer Systems Interface
SCTE	Serial Clock Transmit External (EIA-232-E)
SDH	Synchronous Digital Hierarchy (ITU-T)
SDLC	Synchronous Data Link Control (IBM)
SDS	Scientific Data Systems
SEC	Securities and Exchange Commission
SECAM	Sequential Color with Memory
SELECT	Select Transponder (RFID)
SELECT_ACKNOWLEDGE	Acknowledge Selection (RFID)
SELECT_SNR	Select Serial Number (RFID)
SF	Superframe Format (T-1)
SFGI	Startup Frame Guard Integer (RFID)
SGML	Standard Generalized Markup Language
SGMP	Simple Gateway Management Protocol (IETF)
SHDSL	Symmetric HDSL
SIF	Signaling Information Field
SIG	Special Interest Group
SIO	Service Information Octet
SIP	Serial Interface Protocol
SIR	Sustained Information Rate (SMDS)
SLA	Service Level Agreement
SLIP	Serial Line Interface Protocol (IETF)
SM	Switching Module
SMAP	System Management Application Part
SMDS	Switched Multimegabit Data Service
SMF	Single Mode Fiber
SMP	Simple Management Protocol
SMP	Switching Module Processor
SMR	Specialized Mobile Radio
SMS	Standard Management System (SS7)

SMTP	Simple Mail Transfer Protocol (IETF)
SNA	Systems Network Architecture (IBM)
SNAP	Subnetwork Access Protocol
SNI	Subscriber Network Interface (SMDS)
SNMP	Simple Network Management Protocol (IETF)
SNP	Sequence Number Protection
SNR	Serial Number (RFID)
SOHO	Small-Office, Home-Office
SONET	Synchronous Optical Network
SPAG	Standards Promotion and Application Group
SPARC	Scalable Performance Architecture
SPE	Synchronous Payload Envelope (SONET)
SPID	Service Profile Identifier (ISDN)
SPM	Self Phase Modulation
SPOC	Single Point of Contact
SPX	Sequenced Packet Exchange (NetWare)
SQL	Structured Query Language
SRB	Source Route Bridging
SRP	Spatial Reuse Protocol
SRS	Stimulated Raman Scattering
SRT	Source Routing Transparent
SS7	Signaling System 7
SSCC	Serial Shipping Container Code
SSL	Secure Socket Layer (IETF)
SSP	Service Switching Point (SS7)
SSR	Secondary Surveillance Radar
SST	Spread Spectrum Transmission
STDM	Statistical Time Division Multiplexing
STM	Synchronous Transfer Mode
STM	Synchronous Transport Module (SDH)
STP	Shielded Twisted Pair
STP	Signal Transfer Point (SS7)
STS	Synchronous Transport Signal (SONET)
STX	Start of Text (BISYNC)
SVC	Signaling Virtual Channel (ATM)
SVC	Switched Virtual Circuit
SXS	Step-by-Step Switching
SYN	Synchronization

SYNTRAN	Synchronous Transmission
TA	Terminal Adapter (ISDN)
TAG	Technical Advisory Group
TASI	Time Assigned Speech Interpolation
TAXI	Transparent Asynchronous Transmitter/Receiver Interface (Physical Layer)
TCAP	Transaction Capabilities Application Part (SS7)
TCM	Time Compression Multiplexing
TCM	Trellis Coding Modulation
TCP	Transmission Control Protocol (IETF)
TDD	Time Division Duplexing
TDM	Time Division Multiplexing
TDMA	Time Division Multiple Access
TDR	Time Domain Reflectometer
TE1	Terminal Equipment type 1 (ISDN capable)
TE2	Terminal Equipment type 2 (non-ISDN capable)
TEI	Terminal Endpoint Identifier (LAPD)
TELRIC	Total Element Long-Run Incremental Cost
TIA	Telecommunications Industry Association
TIRIS	TI RF Identification Systems (Texas Instruments)
TIRKS	Trunk Integrated Record Keeping System
TL1	Transaction Language 1
TLAN	Transparent LAN
TM	Terminal Multiplexer
TMN	Telecommunications Management Network
TMS	Time-Multiplexed Switch
TOH	Transport Overhead (SONET)
TOP	Technical and Office Protocol
TOS	Type of Service (IP)
TP	Twisted Pair
TR	Token Ring
TRA	Traffic Routing Administration
TSI	Time Slot Interchange
TSLRIC	Total Service Long-Run Incremental Cost
TSO	Terminating Screening Office
TSO	Time-Sharing Option (IBM)

TSR	Terminate and Stay Resident
TSS	Telecommunication Standardization Sector (ITU-T)
TST	Time-Space-Time Switching
TSTS	Time-Space-Time-Space Switching
TTL	Time to Live
TU	Tributary Unit (SDH)
TUG	Tributary Unit Group (SDH)
TUP	Telephone User Part (SS7)
UA	Unnumbered Acknowledgment (HDLC)
UART	Universal Asynchronous Receiver Transmitter
UBR	Unspecified Bit Rate (ATM)
UCC	Uniform Code Council
UDI	Unrestricted Digital Information (ISDN)
UDP	User Datagram Protocol (IETF)
UHF	Ultra High Frequency
UI	Unnumbered Information (HDLC)
UNI	User-to-Network Interface (ATM, FR)
UNIT™	Unified Network Interface Technology™ (Ocular)
UNMA	Unified Network Management Architecture
UNSELECT	Unselect Transponder (RFID)
UPC	Universal Product Code
UPS	Uninterruptable Power Supply
UPSR	Unidirectional Path Switched Ring
UPT	Universal Personal Telecommunications
URL	Uniform Resource Locator
USART	Universal Synchronous Asynchronous Receiver Transmitter
USB	Universal Serial Bus
UTC	Coordinated Universal Time
UTP	Unshielded Twisted Pair (Physical Layer)
UUCP	UNIX-UNIX Copy
VAN	Value-Added Network
VAX	Virtual Address Extension (DEC)
vBNS	Very High Speed Backbone Network Service
VBR	Variable Bit Rate (ATM)
VBR-NRT	Variable Bit Rate-Non-Real-Time (ATM)

VBR-RT	Variable Bit Rate-Real-Time (ATM)
VC	Venture Capital
VC	Virtual Channel (ATM)
VC	Virtual Circuit (PSN)
VC	Virtual Container (SDH)
VCC	Virtual Channel Connection (ATM)
VCI	Virtual Channel Identifier (ATM)
VCSEL	Vertical Cavity Surface Emitting Laser
VDSL	Very High bit rate Digital Subscriber Line
VDSL	Very High-speed Digital Subscriber Line
VERONICA	Very Easy Rodent-Oriented Netwide Index to Computerized Archives (Internet)
VGA	Variable Graphics Array
VHF	Very High Frequency
VHS	Video Home System
VID	VLAN ID
VIN	Vehicle Identification Number
VINES	Virtual Networking System (Banyan)
VIP	VINES Internet Protocol
VLAN	Virtual LAN
VLF	Very Low Frequency
VLR	Visitor Location Register (Wireless)
VLSI	Very Large Scale Integration
VM	Virtual Machine (IBM)
VM	Virtual Memory
VMS	Virtual Memory System (DEC)
VOD	Video-on-Demand
VP	Virtual Path
VPC	Virtual Path Connection
VPI	Virtual Path Identifier
VPN	Virtual Private Network
VR	Virtual Reality
VSAT	Very Small Aperture Terminal
VSB	Vestigial Sideband
VSELP	Vector-Sum Excited Linear Prediction
VT	Virtual Tributary
VTAM	Virtual Telecommunications Access Method (SNA)

VTOA	Voice and Telephony over ATM
VTP	Virtual Terminal Protocol (ISO)
WACK	Wait Acknowledgment (BISYNC)
WACS	Wireless Access Communications System
WAIS	Wide Area Information Server (IETF)
WAN	Wide Area Network
WAP	Wireless Application Protocol (Wrong Approach to Portability)
WARC	World Administrative Radio Conference
WATS	Wide Area Telecommunications Service
WDM	Wavelength Division Multiplexing
WIN	Wireless In-building Network
WISP	Wireless ISP
WLAN	Wireless Local Area Network
WTO	World Trade Organization
WWW	World Wide Web (IETF)
WYSIWYG	What You See Is What You Get
xDSL	x-Type Digital Subscriber Line
XID	Exchange Identification (HDLC)
XML	Extensible Markup Language
XNS	Xerox Network Systems
XPM	Cross Phase Modulation
ZBTSI	Zero Byte Time Slot Interchange
ZCS	Zero Code Suppression

APPENDIX B

GLOSSARY OF TERMS

3G 3G systems will provide access to a wide range of telecommunication services supported by both fixed telecommunication networks and other services specific to mobile users. A range of mobile terminal types will be supported, and may be designed for mobile or fixed use. Key features of 3G systems are compatibility of services, small terminals with worldwide roaming capability, Internet and other multimedia applications, high bandwidth, and a wide range of services and terminals.

4G 4G networks extend 3G network capacity by an order of magnitude, rely entirely on a packet infrastructure, use network elements that are 100 percent digital, and offer extremely high bandwidth.

A

AAL ATM Adaptation Layer; in ATM, the layer responsible for matching the payload being transported to a requested quality of service level by assigning an ALL Type which the network responds to.

Abend A contraction of the words *abnormal end* used to describe a computer crash in the mainframe world.

Absorption A form of optical attenuation in which optical energy is converted into an alternative form, often heat. Often caused by impurities in the fiber, hydroxyl absorption is the best-known form.

Acceptance Angle The critical angle within which incident light is totally internally reflected inside the core of an optical fiber.

Access The set of technologies used to reach the network by a user.

Accounts Payable Amounts owed to suppliers and vendors for products and/or services that have been delivered on credit. Most accounts payable agreements call for the credit to be reconciled within 30 to 60 days.

Accounts Receivable Money that is owed to the corporation.

ADM Add-Drop Multiplexer; a device used in SONET and SDH systems that has the ability to add and remove signal components without having to demultiplex the entire transmitted transmission stream, a significant advantage over legacy multiplexing systems such as DS3.

Aerial Plant Transmission equipment (including media, amplifiers, splice cases, and so on) that is suspended in the air between poles.

ALOHA The name given to the first LAN, designed and implemented in Hawaii and used to interconnect the various campuses of the state's university system.

Alternate Mark Inversion The encoding scheme used in T1. Every other "one" is inverted in polarity from the one that preceded or follows it.

ALU Arithmetic Logic Unit; the brain of a CPU chip.

Amplifier (1) A device that increases the transmitted power of a signal. Amplifiers are typically spaced at carefully selected intervals along a transmission span. (2) A device used in analog networks to strengthen data signals.

Amplitude Modulation (1) A signal encoding technique in which the amplitude of the carrier is modified according to the behavior of the signal that it is transporting. (2) The process of causing an electromagnetic wave to carry information by changing or modulating the amplitude or loudness of the wave.

AMPS Advanced Mobile Phone Service; the modern analog cellular network.

Analog (1) A signal that is continuously varying in time. Functionally, the opposite of digital. (2) A word that means constantly varying in time.

Angular Misalignment The reason for loss that occurs at the fiber ingress point. If the light source is improperly aligned with the fiber's core, some of the incident light will be lost, leading to reduced signal strength.

APD Avalanche Photodiode; an optical semiconductor receiver that has the ability to amplify weak, received optical signals by multiplying the number of received photons to intensify the strength of the received signal. APDs are used in transmission systems where receiver sensitivity is a critical issue.

Armor The rigid protective coating on some fiber cables that protects them from crushing and from chewing by rodents.

ARPU Average Revenue per User; the average amount of revenue generated by each customer, calculated by dividing total revenue by the total number of subscribers.

ASCII American Standards Code for Information Interchange; a 7-bit data encoding scheme.

ASIC Application-Specific Integrated Circuit, which is a specially designed IC created for a specific application.

Asset What the company owns.

Asynchronous Data that is transmitted between two devices that do not share a common clock source.

ATM Asynchronous Transfer Mode; (1) A standard for switching and multiplexing that relies on the transport of fixed-size data entities called cells which are 53 octets in length. ATM has enjoyed a great deal of attention lately because its internal workings allow it to provide quality of service (QoS), a much-demanded option in modern data networks. (2) One of the family of so-called fast packet technologies characterized by low error rates, high speed, and low cost. ATM is designed to connect seamlessly with SONET and SDH.

Attenuation The reduction in signal strength in optical fiber that results from absorption and scattering effects.

Axis The center line of an optical fiber.

B

Back Scattering The problem that occurs when light is scattered backward into the transmitter of an optical system. This impairment is analogous to echo which occurs in copper-based systems.

Balance Sheet The balance sheet provides a view of what a company owns (its assets) and what it owes to creditors (its liabilities). The assets always equal the sum of the liabilities and shareholder equity. Liabilities represent obligations the firm has against its own assets. Accounts payable, for example, represent funds owed to someone or to another company that is outside the corporation, but that are balanced by some service or physical asset that has been provided to the company.

Bandwidth (1) A measure of the number of bits per second that can be transmitted down a channel. (2) The range of frequencies within which a transmission system operates.

Barcode A machine-scannable product identification label, comprising a pattern of alternating thick and thin lines that uniquely identify the product to which they are affixed.

Baseband In signaling, any technique that uses digital signal representation.

Baud The *signaling rate* of a transmission system. This is one of the most misunderstood terms in all of telecommunications.

Often used synonymously with bits per second, baud usually has a very different meaning. By using multibit encoding techniques, a single signal can simultaneously represent multiple bits. Thus the bit rate can be many times the signaling rate.

Beam Splitter An optical device used to direct a single signal in multiple directions through the use of a partially reflective mirror or some form of an optical filter.

BECN Backward Explicit Congestion Notification; a bit used in frame relay for notifying a device that it is transmitting too much information into the network and is therefore in violation of its service agreement with the switch.

Bend Radius The maximum degree to which a fiber can be bent before serious signal loss or fiber breakage occurs. Bend radius is one of the functional characteristics of most fiber products.

Bending Loss Loss that occurs when a fiber is bent far enough that its maximum allowable bend radius is exceeded. In this case, some of the light escapes from the waveguide resulting in signal degradation.

Bidirectional A system that is capable of transmitting simultaneously in both directions.

Binary A counting scheme that uses Base 2.

Bit Rate Bits per second.

BITS Building Integrated Timing Supply; the central office device that receives the clock signal from GPS or another source and feeds it to the devices in the office it controls.

Bluetooth An open wireless standard designed to operate at a gross transmission level of 1 Mbps. Bluetooth is being positioned as a connectivity standard for personal area networks.

Bragg Grating A device that relies on the formation of interference patterns to filter specific wavelengths of light from a transmitted signal. In optical systems, Bragg Gratings are usually created by wrapping a grating of the correct size around a piece of fiber that has been made photosensitive. The fiber is then exposed to strong ultraviolet light which passes through the grating, forming areas of high and low refractive indices. Bragg Gratings (or filters, as they are often called) are used for selecting certain wavelengths of a transmitted signal, and are often used in optical switches, DWDM systems, and tunable lasers.

Broadband Historically, broadband meant "any signal that is faster than the ISDN Primary Rate (T1 or E1)." Today, it means "big pipe"—in other words, a very high transmission speed.

Broadband In signaling the term means analog; in data transmission it means "big pipe" (high bandwidth).

BSRF Bell System Reference Frequency; in the early days of the Bell System, a single timing source in the Midwest provided a timing signal for all central office equipment in the country. This signal, delivered from a very expensive cesium clock source, was known as the BSRF. Today, GPS is used as the main reference clock source.

Buffer A coating that surrounds optical fiber in a cable and offers protection from water, abrasion, etc.

Bull's-Eye Code The earliest form of bar code, comprising a series of concentric circles so that the code could be read from any angle.

Bundling A product sales strategy in which multiple services (voice, video, entertainment, Internet, wireless, etc.) are sold as a converged package and invoiced with a single, easy-to-understand bill.

Bus The parallel cable that interconnects the components of a computer.

Butt Splice A technique in which two fibers are joined end-to-end by fusing them with heat or optical cement.

C

Cable An assembly made up of multiple optical or electrical conductors, as well as other inclusions such as strength members, waterproofing materials, armor, etc.

Cable Assembly A complete optical cable that includes the fiber itself and terminators on each end to make it capable of attaching to a transmission or receive device.

Cable Plant The entire collection of transmission equipment in a system, including the signal emitters, the transport media, the switching and multiplexing equipment, and the receive devices.

Cable Vault The subterranean room in a central office where cables enter and leave the building.

Call Center A room in which operators receive calls from customers.

Capacitance An electrical phenomenon by which an electric charge is stored in a circuit.

Capacitive Coupling The transfer of electromagnetic energy from one circuit to another through mutual capacitance, which is nothing more than the ability of a surface to store an electric charge. Capacitance is simply a measure of the electrical storage capacity between the circuits. Similar to the inductive coupling phenomenon described earlier, capacitive coupling can be both intentional and unplanned.

CAPEX Capital Expenditures; wealth in the form of money or property, typically accumulated in a business by a person, partnership, or corporation. In most cases capital expenditures can be amortized over a period of several years, most commonly five.

Capital Intensity A measure that has begun to appear as a valid measure of financial performance for large telecom operators. It is calculated by dividing capital spending (CAPEX) by revenue.

Cash Burn A term that became a part of the common lexicon during the dot-com years. It refers to the rate at which companies consume their available cash.

Cash Flow One of the most common measures of valuation for public and private companies. True cash flow is exactly that —a measure of the cash that flows through a company during some defined time period after factoring out all fixed expenses. In many cases cash flow is equated to EBITDA. Usually, cash flow is defined as income after taxes minus preferred dividends plus depreciation and amortization.

CCITT Consultative Committee on International Telegraphy and Telephony; now defunct and replaced by the ITU-TSS.

CDMA Code Division Multiple Access; one of several digital cellular access schemes. CDMA relies on frequency hopping and noise modulation to encode conversations.

Cell The standard protocol data unit in ATM networks. It comprises a five-byte header and a 48-octet payload field.

Cellular Telephony The wireless telephony system characterized by the following: low-power cells; frequency reuse; handoff; central administration.

Center Wavelength The central operating wavelength of a laser used for data transmission.

Central Office A building that houses shared telephony equipment such as switches, multiplexers, and cable distribution hardware.

CES Circuit Emulation Service; in ATM, a service that emulates private line service by modifying the number of cells transmitted per second and the number of bytes of data contained in the payload of each cell.

Chained Layers The lower three layers of the OSI Model that provide for connectivity.

Chirp A problem that occurs in laser diodes when the center wavelength shifts momentarily during the transmission of a single pulse. Chirp is due to instability of the laser itself.

Chromatic Dispersion Because the wavelength of transmitted light determines its propagation speed in an optical fiber, different wavelengths of light will travel at different speeds during transmission. As a result, the multi-wavelength pulse will tend to "spread out" during transmission, causing difficulties for the receive device. Material dispersion, waveguide dispersion and profile dispersion all contribute to the problem.

CIR Committed Information Rate; the volume of data that a frame relay provider absolutely guarantees it will transport for a customer.

Cladding The fused silica coating that surrounds the core of an optical fiber. It typically has a different index of refraction than the core, causing light that escapes from the core into the cladding to be refracted back into the core.

CLEC Competitive Local Exchange Carrier; a small telephone company that competes with the incumbent player in its own marketplace.

CLP Cell Loss Priority; in ATM, a rudimentary single-bit field used to assign priority to transported payloads.

Close-Coupling Smart Card A card that is defined by extremely short read ranges and are in fact similar to contact-based smart cards. These devices are designed to be used with an insertion reader, similar to what is often seen in modern hotel room doors.

CMOS Complimentary Metal Oxide Semiconductor; a form of integrated circuit technology that is typically used in low-speed and low-power applications.

Coating The plastic substance that covers the cladding of an optical fiber. It is used to prevent damage to the fiber itself through abrasion.

Coherent A form of emitted light in which all the rays of the transmitted light align themselves the same transmission axis, resulting in a narrow, tightly focused beam. Lasers emit coherent light.

Compression The process of reducing the size of a transmitted file without losing the integrity of the content by eliminating redundant information prior to transmitting or storing.

Concatenation The technique used in SONET and SDH in which multiple payloads are "ganged" together to form a superrate frame capable of transporting payloads greater in size than the basic transmission speed of the system. Thus, an OC-12c provides 622.08 Mbps of total bandwidth, as opposed to an OC-12, which also offers 622.08 Mbps, but in increments of OC-1 (51.84 Mbps).

Conditioning The process of doctoring a dedicated circuit to eliminate the known and predictable results of distortion.

Congestion The condition that results when traffic arrives faster than it can be processed by a server.

Connectivity The process of providing electrical transport of data.

Connector A device, usually mechanical, used to connect a fiber to a transmit or receive device or to bond two fibers.

Core (1) The central portion of an optical fiber that provides the primary transmission path for an optical signal. It usually has a higher index of refraction than the cladding. (2) The central high-speed transport region of the network.

COT Central Office Terminal; in loop carrier systems, the device located in the central office that provides multiplexing and demultiplexing services. It is connected to the remote terminal.

Counter-Rotating Ring A form of transmission system that comprises two rings operating in opposite directions. Typically, one ring serves as the active path while the other serves as the protect or backup path.

CPU Central Processing Unit; literally the chipset in a computer that provides the intelligence.

CRC Cyclic Redundancy Check; a mathematical technique for checking the integrity of the bits in a transmitted file.

Critical Angle The angle at which total internal reflection occurs.

CRM Customer Relationship Management; a technique for managing the relationship between a service provider and a customer through the discrete management of knowledge about the customer.

CRS Cell Relay Service; in ATM, the most primitive service offered by service providers, consisting of nothing more than raw bit transport with no assigned AAL types.

CSMA/CD Carrier Sense, Multiple Access with Collision Detection; the medium access scheme used in Ethernet LANs and characterized by an "if it feels good, do it" approach.

Current Assets Those assets on the balance sheet that are typically expected to be converted to cash within a year of the publication date of the balance sheet. Current assets typically include such line items as Accounts Receivable, Cash, Inventories and Supplies, any Marketable Securities held by the corporation, Prepaid Expenses, and a variety of other less critical items that typically fall into the Other line item.

Current Liabilities Obligations that must be repaid within a year.

Current Ratio Calculated by dividing the current assets for a financial period by the current liabilities for the same period. Be careful: a climbing current ratio might be a good indicator of improving financial performance, but could also indicate that warehoused product volumes are climbing.

Cutoff Wavelength The wavelength below which single mode fiber ceases to be single mode.

Cylinder A stack of tracks to which data can be logically written on a hard drive.

D

Dark Fiber Optical fiber that is sometimes leased to a client that is not connected to a transmitter or receiver. In a dark fiber installation, it is the customer's responsibility to terminate the fiber.

DAS Direct Attached Storage; a storage option in which the storage media (hard drives, CDs, etc.) are either integral to the server (internally mounted) or are directly connected to one of the servers.

Data Raw, unprocessed zeroes and ones.

Data Communications The science of moving data between two or more communicating devices.

Data Mining A technique in which enterprises extract information about customer behavior by analyzing data contained in their stored transaction records.

Datagram The service provided by a connectionless network. Often said to be unreliable, this service makes no guarantees with regard to latency or sequentiality.

DCE Data Circuit Terminating Equipment; a modem or other device that delineates the end of the service provider's circuit.

DCF Dispersion Compensating Fiber; a segment of fiber that exhibits the opposite dispersion effect of the fiber to which it is coupled. DCF is used to counteract the dispersion of the other fiber.

DE Discard Eligibility bit; a primitive single-bit technique for prioritizing traffic that is to be transmitted.

Debt to Equity Ratio Calculated by dividing the total debt for a particular fiscal year by the total shareholder equity for the same financial period.

Decibel (dB) A logarithmic measure of the strength of a transmitted signal. Because it is a logarithmic measure, a 20 dB loss would indicate that the received signal is one one-hundredth its original strength.

Detector An optical receive device that converts an optical signal into an electrical signal so that it can be handed off to a switch, router, multiplexer, or other electrical transmission device. These devices are usually either NPN or APDs.

Diameter Mismatch Loss Loss that occurs when the diameter of a light emitter and the diameter of the ingress fiber's core are dramatically different.

Dichroic Filter A filter that transmits light in a wavelength-specific fashion, reflecting non-selected wavelengths.

Dielectric A substance that is non-conducting.

Diffraction Grating A grid of closely spaced lines that are used to selectively direct specific wavelengths of light as required.

Digital (1) A signal characterized by discrete states. The opposite of analog. (2) Literally, discrete.

Digital Hierarchy In North America, the multiplexing hierarchy that allows 64 Kbps DS-0 signals to be combined to form DS-3 signals for high bit rate transport.

Diode A semiconductor device that only allows current to flow in a single direction.

Dispersion The spreading of a light signal over time that results from modal or chromatic inefficiencies in the fiber.

Distortion A known and measurable (and therefore correctable) impairment on transmission facilities.

Dopant Substances used to lower the refractive index of the silica used in optical fiber.

DS-0 Digital signal level 0, a 64 Kbps signal.

DS-1 Digital signal level 1, a 1.544 Mbps signal.

DS-2 Digital signal level 2, a 6.312 Mbps signal.

DS-3 A 44.736 Mbps signal format found in the North American Digital Hierarchy.

DSF Dispersion-Shifted Fiber; a form of optical fiber that is designed to exhibit zero dispersion within the C-Band (1550 nm). DSF does not work well for DWDM because of Four Wave Mixing problems; Non-Zero Dispersion Shifted Fiber is used instead.

DSL Digital Subscriber Line; a technique for transporting high-speed digital data across the analog local loop while (in some cases) transporting voice simultaneously.

DSLAM Digital Subscriber Line Access Multiplexer; the multiplexer in the central office that receives voice and data signals on separate channels, relaying voice to the local switch and data to a router elsewhere in the office.

DTE Data Terminal Equipment; user equipment that is connected to a DCE.

DTMF Dual-Tone, Multi-Frequency; the set of tones used in modern phones to signal dialed digits to the switch. Each button triggers a pair of tones.

Duopoly The current regulatory model for cellular systems; two providers are assigned to each market. One is the wireline provider (typically the local ILEC), the other an independent provider.

DWDM Dense Wavelength Division Multiplexing; a form of frequency division multiplexing in which multiple wavelengths of light are transmitted across the same optical fiber. These DWDM systems typically operate in the so-called L-Band (1625 nm) and have channels that are spaced between 50 and 100 GHz apart. Newly announced products may dramatically reduce this spacing.

E

E1 The 2.048 Mbps transmission standard found in Europe and other parts of the world. It is analogous to the North American T1.

EBCDIC Extended Binary Coded Decimal Interchange Code; an 8-bit data encoding scheme.

EBITDA Earnings Before Interest, Tax, Depreciation, and Amortization; sometimes called *operating cash flow,* EBITDA is used to evaluate a firm's operating profitability before subtracting non-operating expenses such as interest and other core, non-business expenses and non-cash charges. Long ago, cable companies and other highly capital-intensive industries substituted EBITDA for traditional cash flow as a *temporary* measure of financial performance without adding in the cost of building new infrastructure. By excluding all interest due on borrowed capital, as well as the inevitable depreciation of assets, EBITDA was seen as a temporary better gauge of potential future performance.

EDFA Erbium-Doped Fiber Amplifier; a form of optical amplifier that uses the element erbium to bring about the amplification process. Erbium has the enviable quality that when struck by light operating at 980 nm, it emits photons in the 1,550 nm range, thus providing agnostic amplification for signals operating in the same transmission window.

Edge The periphery of the network where aggregation, QoS, and IP implementation take place. This is also where most of the intelligence in the network resides.

EDGE Enhanced Data for Global Evolution; a 384 Kbps enhancement to GSM.

Edge-Emitting Diode A diode that emits light from the edge of the device rather than the surface, resulting in a more coherent and directed beam of light.

Effective Area The cross-section of a single-mode fiber that carries the optical signal.

EIR Excess Information Rate; the amount of data that is being transmitted by a user ABOVE the CIR in frame relay.

Encryption The process of modifying a text or image file to prevent unauthorized users from viewing the content.

End-to-End Layers The upper four layers of the OSI Model that provide interoperability.

EPS Earnings per Share; calculated by dividing annual earnings by the total number of outstanding shares.

ERP Enterprise Resource Planning; a technique for managing customer interactions through data mining, knowledge management and customer relationship management (CRM).

ESF Extended Superframe; the framing technique used in modern T-carrier systems that provides a dedicated data channel for non-intrusive testing of customer facilities.

Ethernet A LAN product developed by Xerox that relies on a CSMA/CD medium access scheme.

Evanescent Wave Light that travels down the inner layer of the cladding instead of down the fiber core.

Extrinsic Loss Loss that occurs at splice points in an optical fiber.

Eye Pattern A measure of the degree to which bit errors are occurring in optical transmission systems. The width of the eyes (eye patterns look like figure eights lying on their sides) indicates the relative bit error rate.

F

Facility A circuit.

Facilities-Based A regulatory term that refers to the requirement that CLECs own their own physical facilities instead of relying on those of the ILEC for service delivery.

Faraday Effect Sometimes called the magneto-optical effect, the Faraday Effect describes the degree to which some materials can cause the polarization angle of incident light to change when placed within a magnetic field that is parallel to the propagation direction.

FASB Financial Accounting Standards Board; the officially recognized entity that establishes standards for accounting organizations to ensure commonality among countries and international accounting organizations.

Fast Ethernet A version of Ethernet that operates at 100 Mbps.

Fast Packet Technologies characterized by low error rates, high speed, and low cost.

FDMA Frequency Division Multiple Access; the access technique used in analog AMPS cellular systems.

FEC Forward Error Correction; an error correction technique that sends enough additional overhead information along with the transmitted data that a receiver can not only detect an error but actually fix it without requesting a resend.

FECN Forward Explicit Congestion Notification; a bit in the header of a frame relay frame that can be used to notify a distant switch that the frame experienced severe congestion on its way to the destination.

Ferrule A rigid or semi-rigid tube that surrounds optical fibers and protects them.

Fiber Channel A set of standards for a serial I/O bus that supports a range of port speeds including 133 Mbps, 266 Mbps, 530 Mbps, 1 Gbps, and soon, 4 Gbps. The standard supports point-to-point connections, switched topologies, and arbitrated loop architecture.

Fiber Grating A segment of photosensitive optical fiber that has been treated with ultraviolet light to create a refractive index within the fiber that varies periodically along its length. It operates analogously to a fiber grating and is used to select specific wavelengths of light for transmission.

Frame A variable size data transport entity.

Frame Relay One of the family of so-called fast packet technologies characterized by low error rates, high speed, and low cost.

FRBS Frame Relay Bearer Service; In ATM, a service that allows frame relay frame to be transported across an ATM network.

Freespace Optics A metro transport technique that uses a narrow unlicensed optical beam to transport high-speed data.

Frequency Modulation The process of causing an electromagnetic wave to carry information by changing or modulating the frequency of the wave.

Frequency-Agile The ability of a receiving or transmitting device to change its frequency in order to take advantage of alternate channels.

Frequency-Division Multiplexing The process of assigning specific frequencies to specific users.

Fresnel Loss The loss that occurs at the interface between the head of the fiber and the light source to which it is attached. At air-glass interfaces, the loss usually equates to about 4 percent.

FTTC Fiber-to-the-Curb; a transmission architecture for service delivery in which a fiber is installed in a neighborhood and terminated at a junction box. From there, coaxial cable or twisted pair can be cross-connected from the O-E converter to the customer premises. If coax is used, the system is called *Hybrid Fiber Coax* (HFC); twisted pair-based systems are called *Switched Digital Video* (SDV).

FTTH Fiber-to-the-Home; similar to FTTC, except that FTTH extends the optical fiber all the way to the customer premises.

Full-Duplex Two-way simultaneous transmission.

Fused Fiber A group of fibers that are fused together so that they will remain in alignment. They are often used in one-to-many distribution systems for the propagation if a single signal to multiple destinations. Fused fiber devices play a key role in *passive optical networking* (PON).

Fusion Splice A splice made by melting the ends of the fibers together.

FWM Four Wave Mixing; the nastiest of the so-called fiber nonlinearities. FWM is commonly seen in DWDM systems and occurs when the closely spaced channels mix and generate the equivalent of optical sidebands. The number of these sidebands can be expressed by the equation $1/2(n^3-n^2)$, where n is the number of original channels in the system. Thus a 16-channel DWDM system will potentially generate 1,920 interfering sidebands!

G

GAAP Generally Accepted Accounting Principles; those commonly recognized accounting practices that ensure financial accounting standardization across multiple global entities.

GDP Gross Domestic Product; the total market value of all the goods and services produced by a nation during a specific period of time.

GEOS Geosynchronous Earth Orbit Satellite; a family of satellites that orbit above the equator at an altitude of 22,300 miles and provide data and voice transport services.

GFC Generic Flow Control; in ATM, the first field in the cell header. It is largely unused except when it is overwritten in NNI cells, in which case it becomes additional space for virtual path addressing.

Gigabit Ethernet A version of Ethernet that operates at 1,000 Mbps.

Go-Back-N A technique for error correction that causes all frames of data to be transmitted again, starting with the errored frame.

Gozinta Goes into.

Gozouta Goes out of.

GPRS General Packet Radio Service; another add-on for GSM networks that is not enjoying a great deal of success in the market yet. Stay tuned.

GPS Global Positioning System; the array of satellites used for radiolocation around the world. In the telephony world, GPS satellites provide an accurate timing signal for synchronizing office equipment.

GRIN Graded Index Fiber; a type of fiber in which the refractive index changes gradually between the central axis of the fiber and the outer layer, instead of abruptly at the core-cladding interface.

Groom and Fill Similar to add-drop, groom and fill refers to the ability to add (fill) and drop (groom) payload components at intermediate locations along a network path.

GSM Global System for Mobile Communications; the wireless access standard used in many parts of the world that offers two-way paging, short messaging, and two-way radio in addition to cellular telephony.

GUI Graphical User Interface; the computer interface characterized by the click, move, drop method of file management.

H

Half-Duplex Two-way transmission, but only one direction at a time.

Haptics The science of providing tactile feedback to a user electronically. Often used in high-end virtual reality systems.

Headend The signal origination point in a cable system.

Header In ATM, the first five bytes of the cell. The header contains information used by the network to route the cell to its ultimate destination. Fields in the cell header include Generic Flow Control, Virtual Path Identifier, Virtual Channel Identifier, Payload Type Identifier, Cell Loss Priority, and Header Error Correction.

HEC Header Error Correction; in ATM, the header field used to recover from bit errors in the header data.

Hop Count A measure of the number of machines a message or packet has to pass through between the source and the destination. Often used as a diagnostic tool.

Hybrid Fiber Coax A transmission system architecture in which a fiber feeder penetrates a service area and is then cross-connected to coaxial cable feeders into the customers' premises.

Hybrid Loop An access facility that uses more than one medium. For example, Hybrid-Fiber Coax (HFC, defined above) or hybrids of fiber and copper twisted pair.

HZ Hertz; a measure of cycles per second in transmission systems.

I

ILEC Incumbent Local Exchange Carrier; an RBOC.

Income Statement The income statement is used to report a corporation's revenues, expenses and net income (profit) for a particular defined time period. Sometimes called a *Profit and Loss (P&L) Statement* or *Statement of Operations*, the income statement charts a company's performance over a period of time. The results are most often reported as *earnings per share* and *diluted earnings per share*. Earnings per share is defined as the proportion of the firm's net income that can be accounted for on a per-share basis of outstanding common stock. It is calculated by subtracting preferred dividends from net income and dividing the result by the number of common shares that are outstanding. Diluted earnings per share, on the other hand, takes into account earned or fully vested stock options that haven't yet been exercised by their owner, and shares that would be created from the conversion of convertible securities into stock.

Index of Refraction A measure of the ratio between the velocity of light in a vacuum and the velocity of the same light in an optical fiber. The refractive index is always greater than one and is denoted "n."

Inductance The property of an electric circuit by which an electromotive force is induced in it by a variation of current flowing through the circuit.

Inductive Coupling The transfer of electromagnetic energy from one circuit to another as a result of the mutual *inductance* between the circuits. Inductive coupling may be intentional, such as in an impedance matcher that matches the impedance of a transmitter or a receiver to an antenna to guarantee maximum power transfer, or it may be unplanned, as in the annoying power line inductive coupling that occasionally takes place in telephone lines, often referred to as crosstalk or hum.

Information Data that has been converted to manipulatable form.

Injection Laser A semiconductor laser (synonym).

Inside Plant Telephony equipment that is outside of the central office.

Intermodulation A fiber nonlinearity that is similar to four-wave mixing, in which the power-dependent refractive index of the transmission medium allows signals to mix and create destructive sidebands.

Interoperability (1) Characterized by the ability to logically share information between two communicating devices and be able to read and understand the data of the other. (2) In SONET and SDH, the ability of devices from different manufacturers to send and receive information to and from each other successfully.

Intrinsic Loss Loss that occurs as the result of physical differences in the two fibers being spliced.

IR Infrared; the region of the spectrum within which most optical transmission systems operate, found between 700 nm and 0.1 mm.

IRU Indefeasible Rights of Use; a long-term capacity lease of a cable. IRUs are identified by channels and available bandwidth and are typically granted for long periods of time.

ISDN Integrated Services Digital Network; a digital local loop technology that offers moderately high bit rates to customers.

Isochronous A word used in timing systems that means that there is constant delay across a network.

ISP Internet Service Provider; a company that offers Internet access.

ITU International Telecommunications Union; a division of the United Nations that is responsible for managing the telecomm standards development and maintenance processes.

ITU-TSS ITU Telecommunications Standardization Sector; the ITU organization responsible for telecommunications standards development.

J

Jacket The protective outer coating of an optical fiber cable. The jacket may be polyethylene, Kevlar©, or metallic.

JPEG Joint Photographic Experts Group; a standards body tasked with developing standards for the compression of still images.

Jumper An optical cable assembly, usually fairly short, that is terminated on both ends with connectors.

K

Knowledge Information that has been acted upon and modified through some form of intuitive human thought process.

Knowledge Management The process of managing all that a company knows about its customers in an intelligent way so that some benefit is attained for both the customer and the service provider.

L

Lambda A single wavelength on a multi-channel DWDM system.

LAN Local Area Network; a small network that has the following characteristics: privately owned; high speed; low error rate; physically small.

LANE LAN Emulation; in ATM, a service that defines the ability to provide bridging services between LANs across an ATM network.

Large Core Fiber Fiber that characteristically has a core diameter of 200 microns or more.

Laser An acronym for Light Amplification by the Stimulated Emission of Radiation. Lasers are used in optical transmission systems because they produce coherent light that is almost purely monochromatic.

LATA Local Access and Transport Area; the geographic area within which an ILEC is allowed to transport traffic. Beyond LATA boundaries the ILEC must hand traffic off to a long-distance carrier.

LD Laser Diode; a diode that produces coherent light when a forward biasing current is applied to it.

LED Light Emitting Diode; a diode that emits incoherent light when a forward bias current is applied to it. LEDs are typically used in shorter distance, lower speed systems.

LEOS Low Earth Orbit Satellite; satellites that orbit pole-to-pole instead of above the equator and offer near-instantaneous response time.

Liability Obligations the firm has against its own assets. Accounts payable, for example, represent funds owed to someone or to another company that is outside the corporation, but that are balanced by some service or physical asset that has been provided to the company.

Lightguide A term that is used synonymously with optical fiber.

Line Sharing A business relationship between an ILEC and a CLEC in which the CLEC provides logical DSL service over the ILEC's physical facilities.

Linewidth The spectrum of wavelengths that make up an optical signal.

Load Coil A device that tunes the local loop to the voiceband.

Local Loop The pair of wires (or digital channel) that runs between the customer's phone (or computer) and the switch in the local central office.

LOH Line Overhead; in SONET, the overhead that is used to manage the network regions between multiplexers.

Long-Term Debt Debt that is typically due beyond the one-year maturity period of short-term debt.

Loose Tube Optical Cable An optical cable assembly in which the fibers within the cable are loosely contained within tubes inside the sheath of the cable. The fibers are able to move within the tube, thus allowing them to adapt and move without damage as the cable is flexed and stretched.

Loss The reduction in signal strength that occurs over distance, usually expressed in decibels.

M

M13 A multiplexer that interfaces between DS-1 and DS-3 systems.

Mainframe A large computer that offers support for very large databases and large numbers of simultaneous sessions.

MAN Metropolitan Area Network; a network, larger than a LAN, that provides high-speed services within a metropolitan area.

Manchester Encoding A data transmission code in which data and clock signals are combined to form a self-synchronizing data stream, in which each represented bit contains a transition at the midpoint of the bit period. The direction of transition determines whether the bit is a 0 or 1.

Market Cap(italization) Market cap is the current market value of all outstanding shares that a company has. It is calculated by multiplying the total number of outstanding shares by the current share price.

Material Dispersion A dispersion effect caused by the fact that different wavelengths of light travel at different speeds through a medium.

MDF Main Distribution Frame; the large iron structure that provides physical support for cable pairs in a central office between the switch and the incoming/outgoing cables.

MDU Multi-Dwelling Unit; a building that houses multiple residence customers such as an apartment building.

Message Switching An older technique that sends entire messages from point to point instead of breaking the message into packets.

Microbend Changes in the physical structure of an optical fiber caused by bending, that can result in light leakage from the fiber.

Midspan Meet In SONET and SDH, the term used to describe interoperability. See also *Interoperability.*

Modal Dispersion See *Multimode Dispersion.*

Mode A single wave that propagates down a fiber. Multimode fiber allows multiple modes to travel, while single mode fiber allows only a single mode to be transmitted.

Modem A term from the words modulate and demodulate. Its job is to make a computer appear to the network like a telephone.

Modulation The process of changing or *modulating* a carrier wave to cause it to carry information.

MON Metropolitan Optical Network; an all-optical network deployed in a metro region.

MPEG Moving Picture Experts Group; a standards body tasked with crafting standards for motion pictures.

MPLS A level three protocol designed to provide quality of service across IP networks without the need for ATM, by assigning QoS labels to packets as they enter the network.

MPOA Multiprotocol over ATM; in ATM, a service that allows IP packets to be routed across an ATM network.

MSVC Metasignaling Virtual Channel; in ATM, a signaling channel that is always on. It is used for the establishment of temporary signaling channels as well as channels for voice and data transport.

MTSO Mobile Telephone Switching Office; a central office with special responsibilities for handling cellular services and the interface between cellular users and the wireline network.

MTU Multitenant Unit; a building that houses multiple enterprise customers such as a high-rise office building.

Multimode Dispersion Sometimes referred to as modal dispersion, multimode dispersion is caused by the fact that different modes take different times to move from the ingress point to the egress point of a fiber, thus resulting in modal spreading.

Multimode Fiber Fiber that has a core diameter of 62.5 microns or greater, wide enough to allow multiple modes of light to be simultaneously transmitted down the fiber.

Multiplexer A device that has the ability to combine multiple inputs into a single output as a way to reduce the requirement for additional transmission facilities.

Mutual Inductance The tendency of a change in the current of one coil to affect the current and voltage in a second coil. When voltage is produced because of a change in current in a coupled coil, the effect is mutual inductance. The voltage always opposes the change in the magnetic field produced by the coupled coil.

N

NA Numerical Aperture; a measure of the ability of a fiber to gather light, NA is also a measure of the maximum angle at which a light source can be from the center axis of a fiber in order to collect light.

NAS Network Attached Storage; an architecture in which a server accesses storage media via a LAN connection. The storage media are connected to another server.

NDSF Non-Dispersion Shifted Fiber; fiber that is designed to operate at the low-dispersion second operational window (1,310 nm).

Net Income Another term for bottom-line profit.

NEXT Near-End Crosstalk; the problem that occurs when an optical signal is reflected back toward the input port from one or more output ports. This problem is sometimes referred to as "isolation directivity."

Noise An unpredictable impairment in networks. It cannot be anticipated; it can only be corrected after the fact.

NZDSF Non-Zero Dispersion-Shifted Fiber; a form of single mode fiber that is designed to operate just outside the 1,550 nm window so that fiber nonlinearities, particularly FWM, are minimized.

O

OAM&P Operations, Administration, Maintenance and Provisioning; the four key areas in modern network management systems. OAM&P was first coined by the Bell System and continues in widespread use today.

OBS Optical Burst Switching; a technique that uses a one-way reservation technique so that a burst of user data, such as a cluster of IP packets, can be sent without having to establish a dedicated path prior to transmission. A control packet is sent first to reserve the wavelength, followed by the traffic burst. As a result, OBS avoids the protracted end-to-end setup delay and also improves the utilization of optical channels for variable-bit-rate services.

OC-n Optical Carrier level n; (1) A measure of bandwidth used in SONET systems. OC-1 is 51.84 Mbps; OC-n is n times 51.84 Mbps. (2) In SONET, the transmission level at which an optical system is operating.

OPEX Operating Expenses; those expenses that must be accounted for in the year in which they are incurred.

Optical Amplifier A device that amplifies an optical signal without first converting it to an electrical signal.

Optical Isolator A device used to selectively block specific wavelengths of light.

OSS Operations Support Systems; another term for OAM&P.

OTDR Optical Time Domain Reflectometer; a device used to detect failures in an optical span by measuring the amount of light reflected back from the air-glass interface at the failure point.

Outside Plant Telephone equipment that is outside of the central office.

Overhead That part of a transmission stream that the network uses to manage and direct the payload to its destination.

P

Packet A variable size entity normally carried inside a frame or cell.

Packet Switching The technique for transmitting packets across a wide area network.

Path Overhead In SONET and SDH, that part of the overhead that is specific to the payload being transported.

Payload In SONET and SDH, the user data that is being transported.

PBX Private Branch Exchange; literally a small telephone switch located on a customer prem. The PBX connects back to the service provider's central office via a collection of high-speed trunks.

PCM Pulse Code Modulation; the encoding scheme used in North America for digitizing voice.

Phase Modulation The process of causing an electromagnetic wave to carry information by changing or modulating the phase of the wave.

Photodetector A device used to detect an incoming optical signal and convert it to an electrical output.

Photodiode A semiconductor that converts light to electricity.

Photon The fundamental unit of light, sometimes referred to as a quantum of electromagnetic energy.

Photonic The optical equivalent of the term electronic.

Pipelining The process of having multiple unacknowledged outstanding messages in a circuit between two communicating devices.

Pixel Contraction of the terms "picture element." The tiny color elements that make up the screen on a computer monitor.

Planar Waveguide A waveguide fabricated from a flat material, such as a sheet of glass, into which are etched fine lines used to conduct optical signals.

Plenum The air handling space in buildings found inside walls, under floors, and above ceilings. The plenum spaces are often used as conduits for optical cables.

Plenum Cable Cable that passes fire retardant tests so that it can legally be used in plenum installations.

Plesiochronous In timing systems, a term that means almost synchronized. It refers to the fact that in SONET and SDH systems, payload components frequently derive from different sources, and therefore may have slightly different phase characteristics.

PMD Polarization Mode Dispersion; the problem that occurs when light waves with different polarization planes in the same fiber travel at different velocities down the fiber.

Pointer In SONET and SDH, a field that is used to indicate the beginning of the transported payload.

Polarization The process of modifying the direction of the magnetic field within a light wave.

Preform The cylindrical mass of highly pure fused silica from which optical fiber is drawn during the manufacturing process. In the industry, the preform is sometimes referred to as a gob.

Private Line A dedicated point-to-point circuit.

Protocol A set of rules that facilitates communications.

Proximity-Coupling Smart Card A card that is designed to be readable at a distance of approximately four-to-ten inches from the reader. These devices are often used for sporting events and other large public gatherings that require access control across a large population of attendees.

PTI Payload Type Identifier; in ATM, a cell header field that is used to identify network congestion and cell type. The first bit indicates whether the cell was generated by the user or by the network, while the second indicates the presence or absence of congestion in user-generated cells, or flow-related OA&M information in cells generated by the network. The third bit is used for service-specific, higher-layer functions in the user-to-network direction, such as to indicate that a cell is the last in a series of cells. From the network to the user, the third bit is used with the second bit to indicate whether the OA&M information refers to segment or end-to-end-related information flow.

Pulse Spreading The widening or spreading out of an optical signal that occurs over distance in a fiber.

Pump Laser The laser that provides the energy used to excite the dopant in an optical amplifier.

PVC Permanent Virtual Circuit; a circuit provisioned in frame relay or ATM that does not change without service order activity by the service provider.

Q

Q.931 The set of standards that define signaling packets in ISDN networks.

Quantize The process of assigning numerical values to the digitized samples created as part of the voice digitization process.

Quick Ratio Calculated by dividing the sum of cash, short-term investments and accounts receivable for a given period by the current liabilities for the same period. It measures the degree of a firm's liquidity.

Quiet Zone The area on either side of the Universal Product Code (UPC) that has no printing.

R

RAM Random Access Memory; the volatile memory used in computers for short-term storage.

Rayleigh Scattering A scattering effect that occurs in optical fiber as the result of fluctuations in silica density or chemical composition. Metal ions in the fiber often cause Rayleigh Scattering.

RBOC Regional Bell Operating Company; today called an ILEC.

Refraction The change in direction that occurs in a light wave as it passes from one medium into another. The most common example is the bending that is often seen to occur when a stick is inserted into water.

Refractive Index A measure of the speed at which light travels through a medium, usually expressed as a ration compared to the speed of the same light in a vacuum.

Regenerative Repeater A device that reconstructs and regenerates a transmitted signal that has been weakened over distance.

Regenerator A device that recreates a degraded digital signal before transmitting it on to its final destination.

Repeater See *Regenerator.*

Retained Earnings Represents the money a company has earned less any dividends it has paid out. This figure does not necessarily equate to cash; more often than not it reflects that amount of money the corporation has reinvested in itself rather than paid out to shareholders as stock dividends.

ROI Return on Investment; the ratio of a company's profits to the amount of capital that has been invested in it. This calculation measures the financial benefit of a particular business activity relative to the costs of engaging in the activity.

The profits used in the calculation of ROI can be calculated before or after taxes and depreciation, and can be defined either as the first year's profit or as the weighted average profit during the lifetime of the entire project. Invested capital, on the other hand, is typically defined as the capital expenditure required for the project's first year of existence. Some companies may include maintenance or recurring costs as part of the invested capital figure, such as software updates.

A word of warning about ROI calculations: Because there are no hard and fast rules about the absolute meanings of profits and invested capital, using ROI as a comparison of companies can be risky because of the danger of comparing apples to tractors, as it were. Be sure that comparative ROI calculations use the same bases for comparison.

ROM Read Only Memory; memory that cannot be erased, often used to store critical files or boot instructions.

RPR Resilient Packet Ring; a ring architecture that comprises multiple nodes that share access to a bi-directional ring. Nodes send data across the ring using a specific MAC protocol created for RPR. The goal of the RPR topology is to interconnect multiple nodes ring architecture that is media-independent for efficiency purposes.

RT Remote Terminal; in loop carrier systems, the multiplexer located in the field. It communicates with the COT.

S

SAN Storage Area Network; a dedicated storage network that provides access to stored content. In a SAN, multiple servers may have access to the same servers.

SBS Stimulated Brillouin Scattering; a fiber nonlinearity that occurs when a light signal traveling down a fiber interacts with acoustic vibrations in the glass matrix (sometimes called photon-phonon interaction), causing light to be scattered or reflected back toward the source.

Scattering The backsplash or reflection of an optical signal that occurs when it is reflected by small inclusions or particles in the fiber.

SDH The abbreviation for Synchronous Digital Hierarchy, the European equivalent of SONET.

Sector A quadrant on a disk drive to which data can be written. Used for locating information on the drive.

SEC Securities and Exchange Commission; the government agency that is responsible for regulation of the securities industry.

Selective Retransmit An error correction technique in which only the errored frames are retransmitted.

Shareholder Equity Claims that shareholders have against the corporation's assets.

Sheath One of the layers of protective coating in an optical fiber cable.

Signaling The techniques used to set up, maintain, and tear down a call.

Simplex One way transmission only.

Slotted ALOHA A variation on ALOHA in which stations transmit at pre-determined times to ensure maximum throughput and minimal collisions.

SMF Single Mode Fiber; the most popular form of fiber today, characterized by the fact that it allows only a single mode of light to propagate down the fiber.

SOH Section Overhead; in SONET systems, the overhead that is used to manage the network regions that occur between repeaters.

Soliton A unique waveform that takes advantage of nonlinearities in the fiber medium, the result of which is a signal that suffers essentially no dispersion effects over long distances. Soliton transmission is an area of significant study at the moment, because of the promise it holds for long-haul transmission systems.

SONET Abbreviation for the Synchronous Optical Network, a multiplexing standard that begins at DS-3 and provides standards-based multiplexing up to gigabit speeds. SONET is widely used in telephone company long-haul transmission systems, and was one of the first widely deployed optical transmission systems.

SPM Self-Phase Modulation; the refractive index of glass is directly related to the power of the transmitted signal. As the power fluctuates, so too does the index of refraction, causing waveform distortion.

Source The emitter of light in an optical transmission system.

SRP Spatial Reuse Protocol; A media-independent MAC layer protocol deployed over two counter-rotating optical rings. The rings provide survivability of data in the event of a failed node or cable by rerouting data over the alternate ring. SRP provides offers efficient bandwidth use by ensuring that packets traverse only the part of the ring necessary to get to the destination node.

SRS Stimulated Raman Scattering; a fiber nonlinearity that occurs when power from short wavelength, high power channels is bled into longer wavelength, lower power channels.

SS7 Signaling System Seven; the current standard for telephony signaling worldwide.

Standards The published rules that govern an industry's activities.

Statement of Cash Flows Illustrates the manner in which the firm generated cash flows (the sources of funds) and the manner in which it employed those cash flows to support ongoing business operations.

Steganography A cryptographic technique in which encrypted information is embedded in the pixel patterns of graphical images. The technique is being closely examined as a way to enforce digital watermarking and digital signature capabilities.

Step Index Fiber Fiber that exhibits a continuous refractive index in the core which then steps at the core-cladding interface.

Store-and-Forward The transmission technique in which data is transmitted to a switch, stored there, examined for errors, examined for address information, and forwarded on to the final destination.

Strength Member The strand within an optical cable that is used to provide tensile strength to the overall assembly. The member is usually composed of steel, fiberglass, or Aramid yarn.

STS-1 Synchronous Transmission Signal Level 1; in SONET systems, the lowest transmission level in the hierarchy. STS is the electrical equivalent of OC.

Supply Chain The process by which products move intelligently from the manufacturer to the end user, assign through a variety of functional entities along the way.

Supply Chain Management The management methodologies involved in the supply chain management process. See also *Supply Chain.*

Surface Emitting Diode A semiconductor that emits light from its surface, resulting in a low power, broad spectrum emission.

SVC Signaling Virtual Channel; (1) In ATM, a temporary signaling channel used to establish paths for the transport of user traffic. (2) A frame relay or ATM technique in which a customer can establish on-demand circuits as required.

Synchronous A term that means that both communicating devices derive their synchronization signal from the same source.

T

T1 The 1.544 Mbps transmission standard in North America.

T3 In the North American Digital Hierarchy, a 44.736 Mbps signal.

Tandem A switch that serves as an interface between other switches and typically does not directly host customers.

TDMA Time Division Multiple Access; a digital technique for cellular access in which customers share access to a frequency on a round-robin, time division basis.

Telecommunications The science of transmitting sound over distance.

Terminal Multiplexer In SONET and SDH systems, a device that is used to distribute payload to or receive payload from user devices at the end of an optical span.

Tight Buffer Cable An optical cable in which the fibers are tightly bound by the surrounding material.

Time-Division Multiplexing The process of assigning timeslots to specific users.

Token Ring A LAN technique, originally developed by IBM, that uses token-passing to control access to the shared infrastructure.

Total Internal Reflection The phenomenon that occurs when light strikes a surface at such an angle that all of the light is reflected back into the transporting medium. In optical fiber, total internal reflection is achieved by keeping the light source and the fiber core oriented along the same axis so that the light that enters the core is reflected back into the core at the core-cladding interface.

Transceiver A device that incorporates both a transmitter and a receiver in the same housing, thus reducing the need for rack space.

Transponder (1) A device that incorporates a transmitter, a receiver, and a multiplexer on a single chassis. (2) A device that receives and transmits radio signals at a predetermined frequency range. After receiving a signal, the transponder rebroadcasts it at a different frequency. Transponders are used in satellite communications and in location (RFID), identification, and navigation systems. In the case of RFID, the transponder is the tag that is affixed to the product.

Treasury Stock Stock that was sold to the public and later repurchased by the company on the open market. It is shown on the balance sheet as a negative number that reflects the cost of the repurchase of the shares rather than the actual market value of the shares. Treasury stock can later be retired or resold to improve earnings-per-share numbers if desired.

Twisted Pair The wire used to interconnect customers to the telephone network.

U

UPS Uninterruptible Power Supply; part of the central office power plant that prevents power outages.

V

Venture Capital Money used to finance new companies or projects, especially those with high earning potential. They are often characterized as being high-risk ventures.

Virtual Channel In ATM, a unidirectional channel between two communicating devices.

Virtual Channel Identifier In ATM, the field that identifies a virtual channel.

Vertical Cavity Surface Emitting Laser A small, highly efficient laser that emits light vertically from the surface of the wafer on which it is made.

Vicinity-Coupling Smart Card A card designed to operate at a read range of up to three or four feet.

Virtual Container In SDH, the technique used to transport sub-rate payloads.

Virtual Private Network A network connection that provides private-like services over a public network.

Virtual Path In ATM, a combination of unidirectional virtual channels that make up a bidirectional channel.

Virtual Path Identifier In ATM, the field that identifies a virtual path.

Virtual Tributary In SONET, the technique used to transport sub-rate payloads.

Voice/Telephony over ATM In ATM, a service used to transport telephony signals across an ATM network.

Voiceband The 300 to 3,300 Hz band used for the transmission of voice traffic.

W

Wide Area Network A network that provides connectivity over a large geographical area.

Waveguide A medium that is designed to conduct light within itself over a significant distance, such as optical fiber.

Waveguide Dispersion A form of chromatic dispersion that occurs when some of the light traveling in the core escapes into the cladding, traveling there at a different speed than the light in the core.

Wavelength The distance between the same points on two consecutive waves in a chain—for example, from the peak of wave one to the peak of wave two. Wavelength is related to frequency by the equation $\lambda = c/f$, where lambda (λ) is the wavelength, c is the speed of light, and f is the frequency of the transmitted signal.

Wavelength Division Multiplexing The process of transmitting multiple wavelengths of light down a fiber.

Window A region within which optical signals are transmitted at specific wavelengths to take advantage of propagation characteristic that occur there, such as minimum loss or dispersion.

Window Size A measure of the number of messages that can be outstanding at any time between two communicating entities.

X, Y, Z

XPM Cross-Phase Modulation; a problem that occurs in optical fiber that results from the nonlinear index of refraction of the silica in the fiber. Because the index of refraction varies according to the strength of the transmitted signal, some signals interact with each other in destructive ways. Cross-Phase Modulation is considered to be a fiber nonlinearity.

Zero Dispersion Wavelength The wavelength at which material and waveguide dispersion cancel each other.

INDEX

Note: Boldface numbers indicate illustrations.

B

C

D

E

F

G

H

M

N

O

P

T

U

V

W

X

Z

ABOUT THE AUTHOR

Steven Shepard is a professional writer and educator who specializes in international telecommunications. Formerly with Hill Associates, now president of Shepard Communications Group, he is the author of a number of well-received technical books including *Telecom Crash Course, Telecom Convergence, SONET/SDH Demystified*, and *Optical Networking Crash Course*. He lives and works in Williston, Vermont.